Contents

Edited by EC Hill
BSc, MSc, FInstPetrol, MIMarEST

Sponsored by

Published by IMarEST
The Institute of Marine Engineering, Science and Technology
80 Coleman Street
London EC2R 5BJ

www.imarest.org

A charity registered in England and Wales
Registered Number 212992

A CIP catalogue record for this book is available from the British
Library

ISBN 1-902536-46-0 √

Introduction

Ian Burrows - Past President, IMarEST

Some time ago, the Institute of Marine Engineers (now IMarEST) began to hear reports of accelerated localised corrosion in ships' bilges. In Europe there were many reports of fuel fouling problems. There were reports of accelerated localised corrosion in tanker cargo tanks, particularly in the new breed of double-hulled tankers. There was a common factor to those reports – MICROBES. The Institute's Technical Committee, headed by Dr Jim Cowley, became concerned and set up a subcommittee to study the problem. This was initially chaired by Richard Stuart of Lloyd's Register, and later by myself. Experts from various disciplines were invited to attend.

The existence of the Microbial Subcommittee was publicised and a seminar was held in London. It became clear that those with problems were reluctant to talk about them. Many of those attending the committee were aware of particular incidents but were unable to talk about them because of commercial confidentiality. Eventually the committee was stood down. However, we had enough evidence to believe that many microbial problems went unrecognised.

The Terms of Reference of the committee included a requirement '... to promulgate information on microbial contamination,' and this book is the result. It is believed to be unique. In 1978, the Council of British Shipping published its 56 page Technical Report TR/069 *Microbial Aspects of Corrosion, Equipment Malfunction and System failure in the Marine Industry*, but this is long since out of print. Many books and papers have been written about microbial problems, but most are by specialists for specialists. This publication has been produced to alert ships' staffs, superintendents and others concerned with shipping, to the nature of microbial shipboard problems and to assist in their recognition.

We thank all our contributors, in particular Mr EC Hill who undertook the demanding task of editing this volume.

Edward C Hill

Ted Hill is managing director and consultant at ECHA Microbiology Ltd, Cardiff, which has its origins in the University of Wales. Ted was senior lecturer in industrial microbiology and also ran a consultancy and testing service from 1965-1983, utilising his pioneering experience in the field of industrial microbiology, particularly in investigations of the biodeterioration of petroleum products. He was chairman of the Institute of Petroleum Microbiology Committee from 1985 to 1990 and again from 1992 to 1998 and was presented an Award of Council by the IP in 1998. He was also a member of the Institute of Marine Engineers Microbiology Committee and has presented several papers at the Institute. Ted has specific responsibility for the company's On-site Test Kit business.

Richard Stuart

Richard was an engineer cadet with BP Shipping and attended Glasgow and South Shields Engineering Technical Colleges. From 1974-83 he served as a seagoing engineer officer and then as a technical engineer and superintendent engineer Assistant. After leaving BP Shipping in early 1986, he held further engineering positions with Alfa-Laval Engineering, Sealink British Ferries and Fisher Controls. He joined Lloyd's Register of Shipping in 1990 as an engineer surveyor within the Fluid Analytical Consultancy Services (FACS) department and in 1996 was promoted to senior engineer surveyor. He is the surveyor in charge of the Lubricant Quality Scan (LQS) section within the Fuel Oil Bunker Analysis and Advisory Services (FOBAS) department. He is responsible for the development of LQS and involved in operational investigations and analytical consultancy support to the marine, offshore industry and industrial sectors.

Graham C Hill

Graham graduated from Southampton University in 1983 with a joint honours degree in biology and oceanography and then spent a brief period employed by the microbiology department at the University of Wales, Cardiff, before helping to establish ECHA Microbiology later that year. His 18 years of practical experience includes considerable world-wide on-site attendance, investigating microbial spoilage and corrosion, particularly in the petroleum and marine industries. Graham is the current chairman of the Institute of Petroleum Microbiology Committee and holds positions on the committees of numerous other scientific and technical organisations. He is director of ECHA's Laboratory Testing and Consultancy Services.

Walter G Guthrie

Educated in Scotland at Dumbarton Academy, Paisley College of Technology and Strathclyde University Walter Guthrie gained his Master's degree in pharmaceutical technology in 1973. Thereafter he followed a career in applied chemistry working initially on pharmaceutical and subsequently agrochemical product development. He moved to The Boots Company in the mid 1970s and in

1980 took up a role in technical support for the company's speciality chemical group dealing mainly with the development of anti-microbial technologies. Following the formation of the Boots MicroCheck Group in 1987, Walter was appointed technical development manager for the group's products with responsibility for technical support, new product and application development and regulatory matters.

Boots MicroCheck was sold in 1995 to BASF, later becoming BASF MicroCheck Ltd with global responsibility for the marketing and technical support for the BASF range of biocides. Walter assumed the role of research and development manager.

Anthony Webster

Tony Webster graduated from Loughborough University with a human biology degree in 1988 and in the same year gained employment as a pollution group technical officer for the London Borough of Southwark, working on a wide variety of environmental health issues. In 1990 he took up a post at the Monitoring and Assessment Research Centre responsible for air quality research on behalf of UNEP and WHO. Tony joined Lloyd's Register (LR) in December 1992 as an environmental scientist, principally to work on air quality measurement and evaluation. In 1996 he completed a part-time MSc in environmental science at Birkbeck College and has since left LR to work for another company in the environmental field.

Ms Lou Baxter

Lou. Baxter has a degree in environmental pollution science and biological science and an MSc in European environmental policy and regulation. After graduation, she worked for a water treatment company for three years as an environmental consultant in the field of water and air hygiene. She has also conducted air and water quality surveys and legionella risk assessments. In March 1998, Lou joined LR as an environmental specialist where she continued their research work on shipboard potable and recreational water quality and air quality.

In 2002 Lou Baxter joined DNV, based in the society's UK Maritime Solutions Unit, providing risk management consultancy services to the maritime industry. As a result of her experience she has a wide-ranging knowledge of marine pollution issues, international environmental law and regulations.

Mrs Jan Colligan

Jan Colligan is a food and hygiene inspector with the United Kingdom Maritime and Coastguard Agency. She is a Member of the Royal Environmental Health Institute of Scotland and an Associate member of the Institution of Occupational Health and Safety.

Under the International Labour Organisation (ILO) Convention No 147 the UK has responsibilities concerning proper standards of living conditions and food hygiene for crews on board ships.

As the MCA hygiene specialist, appointed in 1995, Jan carries out regular ship inspection throughout the UK to ensure that satisfactory standards of hygiene and other aspects relevant to living conditions are applied according to Merchant Shipping legislation. She is actively involved in promoting standards within the Agency and with other organisations and represents the MCA on all matters concerning health and hygiene.

David Robbins

David Robbins is a principal member of Dr JH Burgoyne & Partners LLP, a firm of consulting engineers and scientists principally engaged in the investigation of fires and explosions and related incidents. Mr Robbins has attended onboard vessels, on behalf of marine clients, in order to provide advice on the cause of heating and microbial spoilage in various agricultural materials. He has provided expert advice for Court on the spoilage of these cargoes and on contamination in a variety of liquid fuels. He has provided advice and supervision during the extinguishment and control of fires in cargo and on spillages of hazardous chemical cargoes.

Mr Robbins qualified with a degree in biology, awarded by Southampton University in 1983 and afterwards joined Echa Microbiology Ltd. He moved to Burgoynes in 1990 and transferred to the company's Singapore office in 1996. Since August 2000 he has run Burgoyne's Wales and West office in Cardiff.

Dr Barry Herbert

Dr Barry Herbert obtained a PhD in microbiology from Imperial College, London, in 1970. He worked at the Shell Research Laboratories in Sittingbourne, Kent, for 25 years and was responsible for investigating microbiological problems associated with the oil industry. This work covered upstream and downstream aspects, the upstream activities relating to problems associated with sulphate reducing bacteria, reservoir souring, and corrosion. He also gave support to Shell's operating companies world-wide.

Since retiring from Shell, Dr Herbert undertakes private consultancy work.

Glossary

Aerobe: An organism which normally grows in the presence of gaseous or dissolved oxygen.

Algae: Photosynthetic, green or red, organisms, related to higher plants; they vary in size from unicellular algae, a few microns in size, to seaweeds many metres in length. During photosynthesis oxygen is evolved.

Anaerobe: An organism which normally grows in the absence of oxygen.

Anode: A positive electrode where metal anions dissolve and electrons accumulate or are dissipated..

Bacteria: Heterogenous group of microscopic organisms, mostly 1-10 microns in length. Different groups can occupy a range of environments; they chemically utilise and change a diverse range of organic (including hydrocarbons) and inorganic substrates.

Biocides: A chemical which kills life; commonly this term is used for killing microbial life.

Biodegradation: The degradation of a substance by living organisms, particularly microbes.

Biofilm: A film of microbes and polymers on a surface or at an interface.

Biostat: A chemical which retards microbial life but does not kill it in the short term.

Biosurfactants: Biological molecules produced by microbes; they lower surface tension between water and oil and/or emulsify them. Frequently produced by hydrocarbon-degrading microbes.

Buffer: A solution that resists change in pH when an acid or alkali is added, or when the solution is diluted.

Cathode: Site where electrons are absorbed.

Eh. (Redox Potential): A measure in mV of the ability of a system or substance to donate electrons (a reducing agent) or accept electrons (an oxidising agent). The former have a positive voltage and the latter a negative voltage. Eh varies according to pH. Anaerobes require very low or negative Eh.

Facultative: Microbes which can grow aerobically or anaerobically.

Fermentation: Microbial degradation not involving oxygen or any other electron acceptor.
The end products are a mixture of partially oxidised and partially reduced substances

Fungi: Microscopic or large, non-photosynthetic organisms, largely composed of filamentous hyphae (branched and unbranched strands). Large fungi (eg mushrooms) have visible fruiting bodies.

Genus: A group of organisms sharing common fundamental characteristics. Each genus contains several species that differ in minor characteristics.

Hydrogen Embrittlement: Occurs in some alloy steels in the proximity of sulphate reducing bacteria.

Inoculate: Introduce living microbes into or onto a system or substrate.

Microbes (also Micro-Organisms): Microscopic living organisms, particularly bacteria, yeasts and moulds, but also algae, protozoa and viruses.

Micron: (Micrometre) 1/1000th of a millimetre. Abbreviated to μ or μm.

Micro-Organisms: See Microbes

Mould: Small fungi, typically forming visible layers (mycelium).

Pasteurise: Heat kill of a high proportion (but not all) microbes. The higher the temperature the shorter the necessary application time, eg milk can be pasteurised at 63°C for 30 minutes or 71°C for 15 seconds.

pH: A measure of alkalinity (pH>7) and acidity (pH<7) which uses a logarithmic scale, ie. pH9 is ten times more alkaline than pH8.

Preservative: A chemical which prevents the growth of any added microbes.

spp: Species (plural)

Sterilise: Kill all microbes by physical or chemical means.

Sulphate Reducing Bacteria (SRB): A group of microbes, essentially anaerobic, which reduce partially oxidised sulphur compounds (sulphite, thiosulphite, thiosulphate, sulphonates) as well as sulphate. Some form resistant spores, some are salt tolerant and some are thermophilic (heat loving). They need organic nutrients, particularly a variety of organic acids, but cannot utilise hydrocarbons. Hydrogen sulphide or its ions are produced and aggressive corrosion pitting of steel occurs. If nitrite, nitrate, or phosphate are present these are reduced in preference to sulphate.

Surfactants: Chemical or biological agents that lower surface tension between water and oil and/or emulsify them.

Viable: Alive and capable of reproduction.

Water Activity (a_w): A measure of water availability corresponding to 1/100 of % Relative Humidity. Most bacteria require an a_w of >0.95, most moulds >0.80 and most yeasts >0.85 (but some >0.65).

1.   Microbes in the Marine Industry

EC Hill

Contents

Microbes in the Marine Industry

1.   The Nature of Microbes or Micro-Organisms

A variety of microscopic living entities exist and are collectively termed microbes or micro-organisms. This term is interchangeable and, therefore, for the sake of uniformity, the word 'microbes' will be used throughout this chapter and the remainder of this publication.

Hundreds of different species of microbes are capable of proliferating in ships or materials supplied to ships. Some basic knowledge of microbes is desirable for an understanding of the phenomena which they produce and to plan logical on-board anti-microbial strategies. In the problems of the marine industry we are involved with three classes of microbes which are very briefly summarised as follows:

1.1   Bacteria

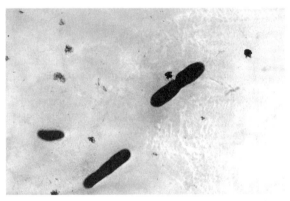

Figure 1. Bacteria illustrating (left to right) how they grow in length and divide into two.

These are spherical, ovoid or, more often, short rods about a micron (1/1000mm) wide and a few microns long. They reproduce simply, by doubling in length and dividing into two and, under ideal growth conditions, they may do this every twenty minutes, rapidly producing enormous microbial populations. Fortunately, although ideal conditions may exist in laboratories, they rarely exist on ships and the time scale for the development of an operational problem is usually several weeks. The progeny may remain loosely attached to each other in clumps or they may separate and disperse. There are thousands of different species. Although in general they prefer neutral or slightly alkaline conditions, some species are not only acid tolerant but they can produce strong mineral acids. Some species produce spores which are very resistant to heat and disinfectants.

1.2 Yeasts

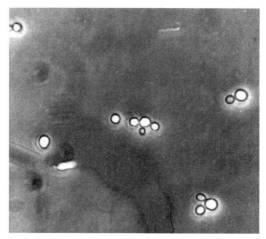

Figure 2. Yeast cells illustrating how they produce 'buds' which grow in size and then separate.

These are typically ovoid, about 5-8 microns long. Reproduction is commonly by buds which form on the parent cell, increase in size and eventually separate. Under some conditions yeasts elongate into substantial filaments of growth and may superficially resemble moulds. The doubling of mass takes several hours even under ideal conditions. Yeasts prefer slight acidity.

1.3 Moulds

These are filamentous organisms, the filament width being a few microns but the length can be many millimetres. The filaments branch and inter-twine and a coherent 'mat' of growth is thus produced. Increase in size is by increase in filament length and by the production of branches; the fastest mass doubling time is several hours but there are limitations due to the rate of nutrient diffusion into the 'mat' and rapid growth may be confined to the periphery. Spores are produced which are inactive but they disperse and germinate to produce new growth 'mats'. Most moulds prefer slightly acid conditions.

1.4 Biomass and Biofilm

Although individual microbes are invisible to the naked eye, their repro-duction will produce visible aggregates and eventually substantial scum and sludge (Biomass) will appear with a tendency to adhere to surfaces (Biofilm).

2.   Naming and Identifying Microbes

2.1   Naming Microbes

Microbes are named in a system which is comparable to that for naming animals and plants. A discreet recognisable type of microbe is a 'species' which has a unique specific name; it is placed within a group with major features in common which has a generic name. The two names are often descriptive, for example the name of a bacterium which causes yellow pus in skin lesions is Staphylococcus aureus. The genus is Staphylococcus (round bacteria in clumps) and the species is aureus (golden colour). However, the naming system is not comprehensive and many industrial microbes have never been named. They may have been allocated an acknowledged 'strain' number instead of a name, but this number may have been assigned and used only in one laboratory. Species may mutate or exchange genetic material and some characteristics therefore vary within the same species. Microbes which cause disease are termed 'pathogenic'; those which are normally spoilage microbes but can cause disease if introduced into the body of a susceptible person are said to be 'opportunistic pathogens'.

2.2   Identifying Microbes

In many marine problems, dozens of different species will be present, each with their own recognisable set of characteristics, and an investigating laboratory may only consider it necessary to give a broad indication of their identity. However, certain species of microbe are known to have particular significance and their presence should be highlighted. Detailed identification becomes important when trying to trace the origins of a problem; for example the source of contamination in a fuel system. A number of criteria such as colour, microscopic appearance and biochemical activity are commonly used as identifying characteristics. Even if they do not yield a previously-allocated species name they can be mathematically processed to yield a unique identifying 'numerical profile' which is sufficient for tracing purposes.

3.   Nutrition and Growth

3.1   Utilisation of Nutrients

Microbes require an aqueous phase for rapid growth as most nutrients diffuse into the cell in aqueous solution. A few microbes have the ability to surround themselves with a hydrated slime which protects them and permits slow growth; for example, some microbes can survive and possibly proliferate slowly in fuel, embedded in the slime which they produce. Moulds can grow slowly in conditions of high humidity.

The microbes must build and repair their cell substances from the nutrients they absorb and hence they require carbon, hydrogen, sulphur, nitrogen and phosphorus in substantial amounts, and lesser amounts of very many other elements. They must also be able to obtain energy from the nutrients to support their vigorous growth processes. This energy is usually derived from the oxidation or fermentation of organic carbon substrates (such as sugars and hydrocarbons) but some organisms can derive energy by the oxidation of inorganic substrates such as nitrites and sulphur or they can use light energy. If molecular oxygen is utilised by microbes for biochemical oxidations they are termed 'aerobic'. Molecular oxygen need not be involved and there are a variety of mechanisms by which one compound is oxidised (oxygen added or electrons removed) whilst another is reduced (oxygen removed or electrons added). Microbes which do not need molecular oxygen are termed 'anaerobic'; for example sulphate reducing bacteria will grow by reducing sulphate (SO_4) to sulphide in the absence of oxygen whilst at the same time oxidising certain carbon compounds. Some microbes can switch between aerobic and anaerobic modes of growth and are termed 'facultative'. Typical facultative microbes are the bacteria which reduce nitrite corrosion inhibitor when cooling water becomes stagnant and deficient in oxygen.

The utilisation of nutrients and chemicals for growth and energy will result in chemical and physical change. Unlike chemical agents of change, microbial agents continue to function almost indefinitely and in most cases the rates of change accelerate. When the physical and chemical environment is not conducive to active growth, the microbes may merely exist passively. In fact, under these 'non-growth' conditions, considerable chemical change can still be catalysed.

Consortia of microbes are involved in the degradation of organic materials and the products of one type may become the food for another type. This is illustrated below.

ORGANIC DEGRADATION

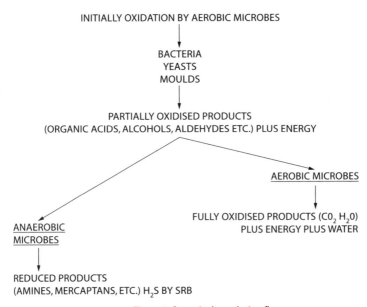

INITIALLY OXIDATION BY AEROBIC MICROBES

BACTERIA
YEASTS
MOULDS

PARTIALLY OXIDISED PRODUCTS
(ORGANIC ACIDS, ALCOHOLS, ALDEHYDES ETC.) PLUS ENERGY

AEROBIC MICROBES

FULLY OXIDISED PRODUCTS (CO_2 H_2O)
PLUS ENERGY PLUS WATER

ANAEROBIC
MICROBES

REDUCED PRODUCTS
(AMINES, MERCAPTANS, ETC.) H_2S BY SRB

Figure 3. Organic degradation flow

In general, the spoilage microbes have characteristics as described above; some have very much simpler nutritive requirements than others.

The pathogenic (disease-causing) microbes are parasites and many require very complex nutrients which they take from the animal body which is their host. Some of the products of their growth will be toxic agents that produce the disease symptoms. Some microbes are termed opportunist pathogens and cause both spoilage and disease. Microbes which cause health problems are of concern onboard and are described in the chapters on water, ventilation and food (Chapters 5, 6 and 7).

Many of the microbiological operational problems that arise in the marine industry are a consequence of the microbial degradation of fuel oil and petroleum products. This may occur during distribution and storage, in end-use and in waste accumulations. Microbial growth is not confined to petroleum products, and fouling and failure of other organic substances, particularly natural fibres, rubber, paints and, of course, food, can be attributed to microbes. Most polymer molecules in plastics are resistant to microbial attack, but the fillers, accelerators and plasticisers are not; the physical characteristics of a formulated plastic may therefore change. For example, flexible PVC may become brittle and porous. It has become increasingly important that coatings applied to steel to prevent corrosion should be tested to ensure that they are resistant to microbial attack.

3.2   Physical Conditions for Growth

Microbes can flourish over a wide range of physical conditions. Some can be found growing slowly in the freezer whilst others occur in hot crude oil tanks and engine coolants. One group can exist at extreme acidity (pH 1), whilst others grow at extreme alkalinity (pH 10). The most abundant growth of a wide variety of species tends to take place from 30-40°C at around neutral pH. In any system there are likely to be temperature gradients, and different populations of microbes may be more active in different parts of the system, according to their temperature optima. Moderate physical pressure (eg hydraulic pressure) has little influence on microbial growth, and moderate salinity (osmotic pressure) has also little influence on growth. Some microbes actually prefer high salt conditions.

It should not, however, be inferred that any one species can flourish over a wide range of physical conditions as each species has its own well-defined set of optimal physical conditions.

4.   Controlling Microbes

Microbes can grow (replicate), remain quiescent, or die; microbes do not die 'naturally' and have to be actively killed. Dead microbes do not disappear and may continue to foul systems.

Although there is a large number of killing mechanisms, only a few can be considered practical on ships. Killing procedures can be considered in two categories; physical and chemical.

4.1 Physical Methods of Killing and Removing Microbes

4.1.1 Heat

Heat killing processes have two variables; temperature and time of application. As an indication of actual practices, milk can be pasteurised at 71°C for 15 seconds or 63°C for 30 minutes. This does not sterilise the milk but considerably reduces the microbial population. We can expect purifier heaters and heated renovating tanks to reduce microbial populations in engine lubricating oil and this will be discussed in detail later.

As a rule of thumb, microbes which flourish at high temperature are more difficult to heat-kill. Hence microbes growing in hot engine lubricants and coolants can be expected to be less sensitive to heat treatment than organisms in milk. Microbes are also known to have different heat sensitivities when suspended in water and when suspended in oil.

4.1.2 Cooling

Cooling retards microbial growth but freezing may actually preserve microbes which are then reactivated when the temperature rises again. Cooling is an impractical method of controlling shipboard microbes, except in food.

4.1.3 Centrifugation

The centrifugal forces generated in a purifier are sufficient to spin out microbes, as they have a specific gravity slightly greater than 1. The rate at which they separate is affected by their degree of aggregation, the viscosity of the suspending oil and the retention time.

4.1.4 Filtration

Microbes are partly removed by conventional filters even though the nominal filter pore size is far greater than the size of the individual microbial cells.

4.1.5 Radiation

Soft (UV) and hard (gamma or electron beam) radiation will kill microbes but there are no practical applications in oils or coolants. UV does not penetrate these adequately but it is an accepted method of treating potable water. Hard irradiation must be contained within too much shielding to be of use. Microwave irradiation produces local heating during food processing and might have the potential for other onboard use. Ultrasound disrupts microbial cells and is being evaluated for decontaminating kerosene. Magnetic devices are being offered commercially for fuel treatment but the scientific basis for their claimed activity is not clear. In some countries, food may be 'preserved' by gamma irradiation.

4.2 Chemical Methods of Controlling Microbes

The terminology which describes anti-microbial chemicals is:

a '.....stat' (eg, bacteriostat inhibits bacterial growth and reproduction but does not kill).
a '.....cide' (eg, fungicide actively kills fungi)

The borderline between the two is usually blurred, as the difference in activity may only be a difference of concentration of the same chemical. In many cases an anti-microbial chemical may actually become a nutrient at very low concentration. The term 'biocide' is loosely used to describe active chemicals. For food-handling hygiene, the equivalent term is 'disinfectant'. Chemicals tend to be active against specific groups of microbes and a selected biocide must have activity against the target microbes. Biocides which exhibit the widest spectrum of activity tend to be the most toxic to humans.

A number of factors determine whether a chemical kills at all, and if it does, the rate at which it does so.

4.2.1 Concentration

Increasing concentration improves performance. Doubling the concentration usually far more than doubles the rate of kill. There are low concentration thresholds involved which determine whether there is any inhibition or kill at all.

4.2.2 Temperature

Raising the temperature of application usually gives a marked short-term improvement in effectiveness. However, many biocides decompose and are inactivated when heated at even moderate temperatures. Hence it is difficult to find a biocide which will persist in hot crankcase oil.

4.2.3 Two-phase Systems

Biocides will partition in oil/water systems according to their relative solubilities in oil and water. It is the final concentration in the aqueous phase which determines the kill rate. This factor is of crucial importance when decontaminating fuel tanks.

4.2.4 Presence of Organic Matter

Most biocides are absorbed by organic matter and their effectiveness is consequently reduced. Even dead bacteria function in this way to reduce biocide activity.

4.2.5   Chemical Deactivation

There are general inactivating mechanisms, anionic surfactants always inactivate for example, cationic biocides, such as quaternary ammonium compounds. There are also specific inactivating reactions between certain other chemicals and biocides. This inactivation may occur rapidly or progressively.

4.2.6   Population Size

More microbes need more biocide.

4.2.7   pH

Biocides have pH optima where they work best and have limits of pH range beyond which they will not work at all and may actually decompose.

The successful use of biocides necessitates selecting the most appropriate anti-microbial product for the circumstances and using it at the correct concentration This concentration must be achieved throughout the system being treated. Additional considerations might be any objectionable smell, ease of handling, and combustibility. Whatever product is used, it must be safe, have an acceptable (or controllable) environmental impact and be in regulatory compliance.

5.   Corrosion

Fouling, detrimental chemical and physical changes and equipment malfunction may be the most obvious indications of microbial proliferation. Microbially-accelerated corrosion may be less visible but the economic consequences are frequently dramatic and expensive and, in some cases, prejudice the safety of the vessel and crew.

5.1   Indirect Mechanisms

Microbes may influence corrosion indirectly, for example by destroying corrosion inhibitors; this is an important corrosive mechanism during the spoilage of coolants and lubricants. They may also destroy protective coatings (see 3.1).

Microbes also directly accelerate normal electrochemical corrosion processes as follows:

5.2   Oxygen Gradients

Most aerobic microbes, when they aggregate in slimes or in crevices, use up oxygen and create an oxygen-deficient zone around them which is anodic in relation to relatively oxygen-rich zones where there are few microbes. Oxygen gradients make electrons flow, and anodic corrosion pits develop under the microbial aggregates.

5.3   Acids

Most microbes produce acids which can be directly corrosive. Weak organic acids are usually produced but a few genera (*Thiobacillus* and *Ferrobacillus*) can oxidise sulphides and sulphur to sulphuric acid. Ferrobacillus can additionally oxidise ferrous compounds to ferric compounds. Sulphuric acid will directly attack steel; this corrosion can occur in crude oil cargo tanks (including humid void space) and in sulphur cargo holds. Organic acids corrode aluminium and bronze; this kind or corrosion can occur in lubricating oil systems. The acidity is usually very localised.

5.4   Depolarisation

Many microbes produce hydrogenase enzymes (catalysts) which remove hydrogen from metal surfaces, thus depolarising them.

5.5   Sulphate Reducing Bacteria (SRB)

These organisms produce the toxic gas, hydrogen sulphide, and ions such as HS^- and S^{2-}, which are highly aggressive to steel and blacken yellow metals. Characteristic craters form in carbon steel but a skeleton of carbon remains which is seen as a graphitic (lead pencil) colour. The bottom of the pit is usually black (ferrous sulphide) although some re-oxidation of this may occur at the surface of the metal where a 'crust' may form. At the same time, SRB hydrogenase enzymes depolarise the steel. When ferrous sulphide forms, it is itself cathodic and continues to drive electron flow and anodic pitting even after the SRB have been killed or have become less active. Some steels progressively become porous and susceptible to hydrogen ingress. Hydrogen embrittlement is known to significantly accelerate stress and fatigue corrosion. Corrosion driven by SRB is very pronounced in oxygen gradients and during intermittent aeration.

Consortia of interdependent aerobic and anaerobic microbes are involved in SRB corrosion. Many different species may be involved in each consortium, and these differ not only from system to system but from point to point in the same system. Micro-environments also exist and differ millimetre by millimetre in terms of pH, oxygen, Eh and chemical composition. They may also change with time, sometimes cyclically. Biofilm is a typical micro-environment for SRB proliferation. The overall microbiological process is usually that oils and occasionally other organic substances first become food for aerobic microbes. Partially-oxidised compounds are formed and become nutrients for other microbes, particularly the SRB. SRB cannot normally feed on hydrocarbons but only on the organic acids, and alcohols produced by the aerobic hydrocarbon degraders. SRB cannot tolerate molecular or dissolved oxygen but they use the oxygen in sulphate (or nitrate) molecules to oxidise organic nutrients. SRB are protected from the inhibitory effect of oxygen by the activity of the aerobic microbes which locally utilise and deplete the dissolved oxygen and, at the same time, change the electrode potential (Eh) from 200-300mV positive to a negative potential; this change is another essential requirement for SRB proliferation. Sulphate in sea water is reduced by SRB to corrosive sulphide. A little of this is assimilated as a nutrient for the reproducing SRB; the remainder disperses as ions or hydrogen sulphide, or reacts with the steel. Many other microbes reduce small amounts of sulphate to sulphide but consume most of it as a nutrient. Other sulphur sources such as sulphurised oil and sulphonates occur in crude oil and oil products and can be reduced to yield hydrogen sulphide.

5.6   Combinations of Mechanisms

The above mechanisms rarely occur in isolation but in various combinations or in succession. Because these corrosion mechanisms are fundamentally chemical or electro-chemical they will be accelerated by a rise in temperature. Corrosion rates are dramatically accelerated by microbes in the field but difficult to reproduce in simulators. Any anti-microbial measure developed in the laboratory must be validated in field situations.

6.   Estimating Numbers of Microbes

There are a number of procedures which can be used to determine the size and/or activity of a microbial population. It may only be necessary to assay a part of the population, for example those microbes which produce corrosive sulphide or those microbes which grow at high temperature. Different microbiological procedures exist and the results of one procedure will not necessarily correspond to results from another procedure. Some degree of interpretation of results may therefore be necessary and, to do this, a knowledge of the principles of the various methodologies is desirable. Not only do the different methodologies deliver different results but also, as microbial contamination is rarely distributed uniformly, the location of the sampling point will markedly influence the level of contamination detected. Any attempt to set microbiological standards or limits should specify both the sampling point and the test method. Some test methods are only suitable for laboratory use and will only be referred to briefly.

Most tests require 'incubation'; cheap incubators are available but in many cases a suitable warm location can be found onboard.

Before proceeding with tests, the nature of the sample must be considered, as different methods are used for oil and water phases. If oil samples contain free water this is best removed and analysed separately using appropriate methods for water phase. The precise procedure for drawing the aqueous sub-sample for analysis should be defined and noted as this can have a dramatic effect on the result obtained (eg water taken from a fuel interface may yield a significantly higher microbial count than water several millimetres below the interface). A recommended procedure is to swirl the sample and then use a sterile pipette to transfer several millilitres of water from just below the interface to a small sterile container. The separated water must be shaken briefly prior to analysis. The remaining oil phase can then be analysed. There are normally many more microbes in the water phase than in the oil phase, by a factor of at least 10^3. Numbers of microbes in lubricating oil or water are usually expressed per ml but are expressed in fuel as per litre.

6.1　Microscope Examination

The method is quick but, realistically, for examining water and sludge it can only be used in a shore-based laboratory by skilled operators using a good microscope. All microbes or microbial fragments, however large, are counted as one unit and they may he dead or alive. Mould fragments can be filtered from fuel onto a membrane (method IP 472/02) and the membrane is then examined microscopically. Portable hand-held microscopes are, in fact, available but their use calls for considerable skill and experience.

6.2　Total Biomass

If sufficient microbial substance is present it may be possible to assess it in a laboratory by chemical/physical methods such as dry weight, total protein, total organic carbon or total organic nitrogen. Biomass assayed could be living or dead.

6.3　Estimates of Numbers of Viable Aerobic Microbes

Many laboratory and on-site tests depend on the ability of viable (living) microbes to reproduce in or on a nutrient gel to yield visible 'colonies' of growth. Counts of these colonies are related directly to the number of viable microbial cells (or aggregates of cells) originally present. The count is referred to as a count of colony forming units (cfu); when numbers are large they may be estimated by comparison to calibration charts. The procedure usually takes one to five days. Non-viable microbes are not assayed. There are standard laboratory procedures which are usually referred to as Total Viable Counts (TVC) and these are usually considered to be the definitive methods.

6.3.1   Aqueous Samples.

6.3.1.1   The common onboard equivalent of a Total Viable Count for aqueous samples is a dip-slide test.

Dip-slides
Description:
  Semi-quantitative test kit for viable aerobic bacteria, yeast and mould cfu in aqueous phase. The test consists of a plastic strip coated on one side with a solid nutrient gel culture medium for bacteria and on the other a solid nutrient gel medium for yeasts and moulds. The strip is mounted on the inside of the screw cap of a sterile plastic tube.
Procedure:
  The strip is removed from its tube and dipped into the aqueous sample or, alternatively, the sample is squirted over both sides of the strip using a sterile pipette; it is then incubated at a temperature close to that of the system sampled for two to five days. The numbers of bacteria, yeast and mould cfu which adhere to the gels are estimated per millilitre of water phase by comparing the numbers of colonies on each side of the dip-slide to reference charts. A growth-indicating dye colours bacterial colonies, red.
Sensitivity:
  100 bacteria cfu per ml and 40 mould and/or yeast cfu per ml of water.
  Analysis Time: up to five days.
Limitations:
  Not quick. Can NOT be used directly on fuel. Underestimates numbers if traces of biocide are present. Calibration is affected by the viscosity of the sample.
Advantages:
  Cheap. Simple to use onboard. Detects most viable microbial contaminants and distinguishes between bacteria, yeasts and moulds. Dip-slides are also used to check the cleanliness of food-handling equipment by pressing them onto a surface.

6.3.1.2   Tests for microbial enzyme activity

There are a number of microbial enzymes which will convert specific chemical substrates into coloured dyes. Several test kits utilise these reactions but only one is suitable for onboard use, the Sig Rapid WB test. The colour change is related to microbial activity rather than microbial numbers.

Sig Rapid WB Test
Description:
  Rapid, enzymatic, semi-quantitative colour test for total microbes in water associated with fuel. A glass tube is supplied which contains a chemical that

will change colour in the presence of certain microbial enzymes.

Procedure:

5ml of water is added to the tube containing the chemical reagent pill; any yellow colour in the water is noted immediately using the colour comparison chart supplied. The test is then incubated at close to body temperature (37°C) for one hour, a colour developer pill is added and the yellow colour read again. Increase in yellow colour is correlated to the extent of microbial contamination in the water phase using the chart.

Sensitivity:

10 000 microbes per ml of water.

Analysis Time:

One hour.

Limitations:

Does not distinguish between different types of microbes. Can only be used to test water associated with fuel. Not particularly sensitive; only detects an existing contamination problem, not a potential problem. Will continue to give a positive result for several days after a biocide has been used.

Advantages:

Quick, simple test.

6.3.1.3 ATP Assays

Adenosine Tri-Phosphate (ATP) is the energy store contained in all living cells including microbes. ATP can be assayed quickly and quantitatively as it promotes the reaction between luciferase and luciferin (extracted from fireflies) producing light which is measured in a dedicated photometer. The technology has not yet been extensively exploited in the marine industry but protocols to extract and test the ATP from samples have now been developed and instruments and reagents are commercially available (HyLiTE Industrial, Merck Ltd) and can be used in the laboratory and onboard for most aqueous samples.

HY-LiTE ATP Test

Description:

Very rapid semi-quantitative test for ATP as an indicator of microbial contamination. ATP occurs in all microbes and is their energy store. A dedicated photometer is supplied to measure the light emitted by the reagent 'pens'.

Procedure:

A 'pen' is dipped into a water sample and an amount is drawn up automatically into the pen. After a simple manipulation of the pen, any ATP present reacts with reagents and causes luminescence. The amount of light emitted is read as Relative Light Units (RLU) by inserting the pen into the special photometer supplied, and this can be correlated to the extent of microbial contamination.

Sensitivity:

At best, about 1000 microbes per millilitre, but this is sufficient for all normal applications.

Analysis Time:

A few minutes.

Limitations:

Requires relatively expensive equipment. Does not distinguish between different types of microbes (bacteria, yeasts and moulds). Recently-killed microbes (eg after biocide treatment) will continue to give an ATP reading. The pens should be refrigerated until use.

Advantages:

Very quick. Detects all viable microbial material. Although usually a laboratory-based test, the photometer is portable and has a battery option for onboard use. This technology is widely used to test swabs of food-handling equipment.

6.3.2   Oil and Fuel Samples

The most commonly used laboratory protocol for distillate fuel is IP 385/99. A measured volume of fuel is filtered through a sterile membrane which retains any microbes present on the membrane surface; the membrane is laid on the surface of a nutritive agar gel and incubated. Any microbes on the membrane absorb nutrients from the gel and reproduce to form visible colonies which can be counted. Although semi-portable equipment is available for this test, it is not usually carried out as an onboard procedure.

Heavy fuel oils and lubricants are too viscous to use this procedure. They can be mixed with a sterile solution of an emulsifier and the emulsion obtained is then tested by a viable count procedure as a laboratory test or semi-quantitatively with a dip-slide as an on-board test. Distillate fuel can also be tested in this way (AFNOR NF M07-070:93).

Two onboard tests for fuel, the HUM Bug and Liquicult tests, are based on the following principle.

Description:

Rubber sealed-bottle of culture 'broth' containing a growth indicator plus syringe with needle.

Procedure:

A known volume of fuel is injected into the broth through the rubber seal. The bottle is shaken and incubated. Growth is indicated by turbidity and/or a colour change.

Sensitivity:

Not specified but depends on the volume of fuel injected; probably 500 – 1000 microbes per millilitre.

Analysis time:
   One to five days.
Limitations:
   This is basically a Go/No Go test although the amount of growth is said to indicate the degree of contamination. The use of hypodermic needles onboard is a possible hazard.

A variant of this method is the Bugbuster test. 1ml of fuel is shaken with sterile water and, after settling, this water is injected with a hypodermic syringe into various bottles of 'broth'. This is basically a Go/No Go test with a sensitivity of about 1000 microbes per litre but it is claimed that heavy fuel contamination causes more bottles to show growth.

The only quantitative onboard test for microbes in fuel and oil is the MicrobMonitor[2] test which utilises thixotropic nutrient gel.

MicrobMonitor[2]
Description:
   On-site, quantitative culture method for viable microbes in fuel or oils, (can also be used for testing water-phase samples but is an expensive alternative to dip-slides).
Procedure:
   A known volume of up to 0.5ml of kerosene or 0.25 ml of gas oil/diesel is added to a bottle of thixotropic nutrient gel which is shaken vigorously to liquefy it and to disperse the sample. It is then tapped to form a horizontal layer of gel which sets solid on standing; the test is incubated for one to five days. Colonies of microbes develop and are counted and then a calculation is made to express the numbers as cfu per litre of fuel. A growth-indicating dye colours colonies purple to aid colony counting.
   When testing lubricant and hydraulic oils, a 0.01ml sample is tested. Measuring devices are included with the kit.
Sensitivity:
   2000 cfu per litre of kerosene, 4000 cfu per litre gas oil/diesel fuel, 100 cfu per ml lubricants or water.
Analysis Time:
   One day (heavy infections) to five days (light or no infection).
Limitations:
   Result is total number of microbes present, as test does not distinguish easily between different types of microbes (bacteria, yeasts and moulds).
Advantages:
   Simple to use on-site. Quantitatively detects viable microbial contaminants. Results correlate well with IP385/99. Faster than laboratory analysis (but not rapid). Can be used to test aqueous samples if dip-slides are not available

6.4   Estimates of Numbers of Viable Anaerobic Microbes

Sulphate reducing bacteria (SRB) are the only microbes in this group which are routinely assayed. The laboratory technique used is sometimes called an end-point dilution count. The sample is diluted ten-fold progressively, and each dilution is inoculated into a liquid culture medium designed to detect SRB. The greater the dilution yielding positive growth, the greater the number of SRB present in the original sample.

There are several tests for SRB which can be used onboard. They all consist of a glass tube of a nutritive gel which has been formulated to grow SRB but they differ in the manner of adding the sample. In the Easicult S test, the sample is added with a glass capillary, in the Sanicheck SRB test the sample is thrust into the gel with a sterile 'pipe cleaner' and in the Sig Sulphide test the sample is poured onto the gel. In each case, the rate and extent of blackening of the gel is equated to the degree of SRB contamination. The tests are designed to test water samples but they can be used less reliably to test oil/fuel samples. Only the Sig Sulphide test will be described.

Sig Sulphide Test
Description:
> Semi-quantitative test for viable SRB in water phase. Used to assess the risk of microbially-induced corrosion in the bottoms fuel or water tanks and in bilges, to investigate the cause of sulphide spoilage of fuel and to detect the potential for generation of toxic hydrogen sulphide gas.

Procedure:
> About 2ml of water is poured or pipetted onto the surface of the gel, or a swab of a corrosion pit stabbed into the gel. The tube is incubated for one to five days and observed regularly for blackening. The rate and extent of blackening is compared with a graph and this indicates the numbers of SRB present. (No calibration is provided for the other SRB tests).

Sensitivity:
> about one SRB per millilitre of water.

Analysis Time:
> one day (heavy infection) to five days (light or no infection).

Limitations:
> Can be erratic if used directly for fuel or oil.

Advantages:
> Cheap. Simple to use on-site. Much faster than standard laboratory methods; results correlate well with standard NACE TMO-194-94 method. Good indicator of severe, well-established anaerobic microbial infections and risk of corrosion.

Some of the methods described above can give erroneous results if there are sub-lethal concentrations of biocide present, as the traces of biocide may slow or halt growth in the test device.

6.5   Assays for Biocides

If biocides are in use they are only effective if they are present at the correct concentration and it is sometimes desirable to check this concentration. The following technique can do this in a laboratory or onboard.

Biocide Rapide Test
This test will only function properly if the tests are incubated in a thermostatic incubator block which can be supplied with the test kits.

Description:
  Bioassay for detection of anti-microbial chemicals (biocides) in water. An ampoule contains a purple nutritive gel and spores of a microbe which is very sensitive to biocides; samples are added to this ampoule which is then incubated at 64°C. The test can be used as a Go/No Go test or used semi-quantitatively if calibrated for specific biocides and a range of sample dilutions is tested. Useful for determining whether fuel, or water associated with fuel, is adequately treated with biocide or whether effluents or waste water contain unacceptable concentrations of biocides or other anti-microbial chemicals. It can only be used to assay biocides in fuel after a standard aqueous extraction procedure.
Procedure:
  Add samples or appropriate dilutions of samples to test ampoules.
  Incubate for three hours at 64°C using an incubator block.
  Note any development of yellow coloration (spore germination and growth = no anti-microbial activity detected). If the purple colour persists, biocide has been detected. Refer to instruction leaflet.
Sensitivity:
  Depends on biocide but most marine biocides are detected. The fuel biocides, Biobor JF, Kathon FP1.5 and Mar 71, can all be detected.
Analysis Time:
  Three to four hours
Limitations:
  For quantitative analysis, the sensitivity of the test to the particular biocide in use needs to be predetermined by testing standard solutions. Needs mains electricity 110/220V for the incubator block.
Advantages:
  Rapid; generally much faster, cheaper and simpler than laboratory chemical analysis.

6.6   Other kits

Other test kits exist; some of them are designed to detect harmful microbes and these are not suitable for use by unqualified staff onboard as they can create a hazard in use. Most test kits rely on growing cultures or colonies of microbes under contained and controlled conditions and they must be disposed of safely according to the supplier's instructions.

7.   Summary

A knowledge of the nature of microbes, the factors which control their growth and death, and the phenomena associated with microbial growth are relevant to an understanding of and to the control of most marine spoilage and corrosion situations and also for creating a healthy environment onboard. The test kits that have been described can be used safely onboard or offshore and give valuable information to facilitate this control.

A variety of niches exist onboard and offshore where microbes can flourish and these are described in the following chapters. In many cases, guidance is given in the selection of the most appropriate microbiological tests or test kits, and reference can be made to this introductory chapter for test details.

8.   Further Reading

1. Hill, EC, *Keeping Shipboard Water Free from Disease*. Marine Engineers Review, December 1988, pp 17-19.
2. *Microbes in Fuels, Lub Oils and Bilges - Recognition and Monitoring. Proc. Seminar*. Institute of Marine Engineers, London 23rd February 1993. 56 pages
3. *Prevention of Microbiological Growth in a Sub-sea Hydraulic System, in Tribology – Solving Friction and Wear Problems*, Technische Akademie, Esslingen, ed. WJ Bartz, pp 2223–2228, 9–11 Jan. 1996
4. Hill, EC and Hill GC, *Microbiological Problems in Distillate Fuels*, Trans. IMarE, 104, (4), pp 119–130. 1992
5. Hill, EC and Hill, GC, *Microbial Proliferation in Bilges and its Relation to Pitting Corrosion of Hull Plate of In-shore Vessels*. Trans. IMarE 105 (4), pp 175–182, 1993
6. Hill, EC. *Fuel Biocides, in A Handbook of Biocide and Preservative Use*, ed HW Rossmore, Blackie Academic and Professional, pp 207–237, 1995
7. Hill, EC and Hill GC, *Microbiological Pitting Corrosion. Old Problems in New Places Mechanisms, Recognition and Control, Proc. Marine Corrosion Prevention*. R Inst. Naval Architects, London 11–12 Oct. 1994
8. Hill, EC and Hill GC. *Microbial Fouling in Ships' Tanks*. Shipbuilding Technology Int'l. ed R Burnett, Sterling publications, pp 179–181, 1994
9. Hill, EC, Collins DJ and Hill GC. *Microbiological Monitoring On-site*. Proc. Int Conf on Microbiological Monitoring On-site, 12–15 April 1999. Univ of Wales, Swansea. Eds. MH Jones and DJ Sleeman, Coxmoor Publ. Co.

2. Microbes in Fuels; Effects on Safety of Ships, and the Health of Crews

RA Stuart

Contents

Introduction

1. Safety of Ships

2. Microbes
2.1 Types of Microbes
2.2 Conditions for Microbes
2.3 Sources of Microbial Contamination
2.3.1 Sea Water
2.3.2 Refinery Practices
2.3.3 Onboard
2.3.4 Cargoes
2.4 Symptoms of Microbial Contamination

3. Effects of Microbial Contamination
3.1 Composition of Fuels
3.2 Microbial Spoilage Process
3.3 Microbial Corrosion Process

4. Prevention and Elimination of Microbial Contamination
4.1 Prevention
4.1.1 Physical Prevention
4.1.2 Chemical Prevention
4.2 Elimination
4.2.1 Physical Decontamination
4.2.2 Chemical Decontamination

5. Health of Crews
5.1 Microbial Hazards
5.2 Chemical Biocides
5.3 Test Kit Disposal

6. Costs to Industry
6.1 Marine Casualties of Microbial Contamination

Introduction

This chapter addresses the microbial problems in fuels being experienced by the marine industry which have reached notable proportions. The industry is now witnessing spoilage, corrosion and operational problems in areas which are least expected, often causing severe damage to hull, machinery and equipment. Shipboard contamination may be initiated from previously infested tanks and systems, or be introduced onboard through contaminated fuel or sea water. Poor housekeeping methods, adverse environmental legislation and poor ship design are all conducive to microbial proliferation and its associated problems

Far from being an 'act of God', microbial damage is almost entirely man-made and preventable; consequently hull insurance is often refused for coastal trade and inshore workboats, unless appropriate anti-microbial preventive procedures are implemented.

Work conducted by The Institute of Marine Engineers (now IMarEST) Microbiological Sub-Committee Members and Lloyd's Register of Shipping examined the ways in which microbial contamination can be reduced by implementing controls, good housekeeping and chemical biocides. To achieve this, new standards are suggested to monitor these measures, thus ensuring that the safe operation of ships is not jeopardised and human health is not endangered.

1.   Safety of Ships

Microbiological attack on distillate fuels seemed to be at its height in the 1970s and early '80s and a large number of ships were affected. As the marine industry found out more about the causes, so the industry was able to effect remedies. Normally, once a cycle has peaked, the problem is virtually over. To some extent this has been true with microbial attack on distillate fuels, but it has not been totally eradicated. While microbial problems in the marine industry were originally mainly confined to distillate fuels, the areas of attack have also spread to light residual fuels.

Curious though this appears on a preliminary examination, there are in fact several reasons for it, as follows:

- Lower levels of shipboard manning have proved disastrous for maintaining stringent housekeeping.
- Adverse trading conditions have led to ships being laid up or in intermittent service, providing long, undisturbed incubation periods for opportunist microbes.
- Marine pollution legislation under the MARPOL 73/78 regulation which restricts the pumping of fuel tank bottom water drains, has led to water laying stagnant for longer periods.
- Environmental restrictions in the use of toxic biocidal chemicals within fuels, exacerbate the problems associated with microbial contamination.
- Design of fuel tanks and pipe systems fail to provide effective water draining and subsequent fuel treatment.

● Lack of knowledge of the factors which cause microbial contamination and lack of accurate diagnosis of the operational problems being experienced.

Consideration of these factors, as well as explaining the increase in spoilage and corrosion problems, also identifies how strategies for recognition, evaluation, rectification and control of microbial infestation may be implemented.

2.   Microbes

Whilst even the worst marine oil spills are eventually broken down by microbes, few of us appreciate that the same microbes are equally content to live in oil onboard. The microbiological contamination process is a well known phenomenon and small populations of microbes exist quite naturally. These microbes are easily tolerated at low contamination levels. It is only when their numbers are not controlled within their immediate environment, that they undergo rapid growth, resulting in a significant infestation.

From a marine point of view, microbiological infestation of fuels may effectively compromise the safety of ships.

2.1   Types of Microbes

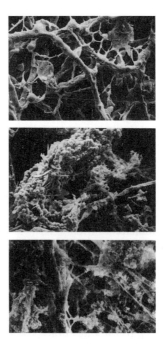

Figure 1. The formation of extensive, thick, tough, intertwined mycelial mats by stages

Figure 2. Filter debris showing rod-shaped bacteria, branched moulds, fungi and yeasts

In the problems of the marine industry, three basic types of spoilage and corrosive microbes are identified; bacteria (including the anaerobic sulphate reducing bacteria, SRB), yeasts and moulds. Their general characteristics have been described in Chapter 1. Those microbes most specific to fuel spoilage are:

Bacteria: typical bacteria known to utilise hydrocarbons are *Pseudomonas aeruginosa*, other *Pseudomonas* species, *Flavobacterium* spp, *Acinetobacter* spp, *Alcaligenes* spp, *Micrococcus* spp, *Arthrobacter* spp, *Corynebacterium* spp, *Brevibacterium* spp, *Klebsiella* spp

Sulphate reducing bacteria: typical SRB which cause sulphide souring of hydrocarbons and sulphide corrosion are *Desulfovibrio* spp, *Desulfotomaculum* spp, *Desulfobulbus* spp.

Yeasts: typical yeasts growing on hydrocarbons are *Candida* spp, *Saccharomyces* spp, *Torula* spp, *Torulopsis* spp, *Hansenula* spp.

Moulds: typical moulds which degrade hydrocarbons are *Hormoconis resinae* spp, *Penicillium* spp, *Aspergillus* spp, *Fusarium* spp, *Monilia* spp, *Botrytis* spp, *Cunninghamella* spp, *Scopulariopsis* spp.

Overall, the microbes which are identified most often in oil in the marine industry are the hydrocarbon degrading spoilage and corrosive species. Their scientific identity is of most interest when trying to trace sources of contamination.

The moulds *Cladosporium resinae*, (now renamed *Hormoconis resinae*), and *Aspergillus fumigatus* tend to form a coherent mat (Figure 1) under which intense corrosion can occur. The aerobic bacteria *Pseudomonas* spp can cause considerable surfactancy problems; they require the presence of dissolved oxygen. Depletion of oxygen then serves to encourage the growth of anaerobic bacteria. The anaerobic bacteria *Desulfovibrio* spp and *Desulfotomaculum* spp will corrode hull, machinery and equipment.

Today, responsibility for spoilage is shared between a very broad spectrum of bacteria, yeasts and moulds, as shown in Figure 2.

2.2   Conditions for Microbes

Microbes are living organisms and their growth depends upon the ready availability of water, nutrients, warmth and oxygen (or sometimes lack of it) within an otherwise acceptable environment.

Water: the main requirement for microbial proliferation is water. This is indicated in the majority of fuel microbial problems reported, which have identified the presence of water within the storage and service tanks due to infrequent draining. Substantial microbial growth needs substantial free water, probably more than 1% water content, since the microbes live in the water phase, but feed off nutrients within the fuel phase.

Nutrients: all microbes require equable conditions for their nourishment and growth. Hydrocarbons and chemical additives in the fuel act as their food source, coupled with those nutrients that are available from polluted harbour water. Rust and other particulates also seem to stimulate microbial growth. As in the cycle of life, dead microbes feed living ones.

Temperature: temperature is a vital factor. Warm conditions encourage growth, while extreme cold below 5°C and excessive heat above 70°C will inhibit the most hardened of microbes. They prefer a temperate climate, in the temperature range of 15°C to 35°C. Warm engine rooms provide ideal breeding grounds.

Environment: microbes dislike undue agitation, preferring the fuel systems to lie dormant. Laid-up ships, or ships in intermittent service, are the most vulnerable to attack. Fuel tanks for long term storage are susceptible to infection.

Ideal conditions for microbial growth do not normally occur in practice, and dry, clean, low temperature fuel will never permit significant growth of microbes. Given the ideal conditions required, significant microbial growth will take several months under shipboard conditions, and severe corrosion, if it occurs, will not appear for many weeks after microbial growth has become apparent.

2.3   Sources of Microbial Contamination

Operational problems due to microbial contamination may be caused by imported infested hydrocarbons or sea water, previous onboard contamination, a combination of both or result from poor onboard operational procedures.

2.3.1   Sea Water

Sea water with negligible pollution will normally contain less than 10^3 bacteria per cm^3 and minimal quantities of yeasts and moulds. Only about 0.1% of these bacteria will be hydrocarbon degrading and SRB are even rarer. Sea water confined in harbours and estuaries with a long history of oil spillage and oil tank and sewage discharges will contain far in excess of 10^3 bacteria per cm^3, including hydrocarbon degraders and large numbers of SRB. Additionally, phosphorus and nitrogen pollutants from agricultural fertilisers, plus corrosion inhibitors and oil additives, will ensure that sufficient nutrients are present to nourish microbes. Harbour and estuary waters may contain up to 11ppm of nitrogen and 2ppm of phosphorus and are far more nutritious to microbes than clean sea water, which would not contain more than 1ppm of nitrogen and phosphorus.

2.3.2   Refinery Practices

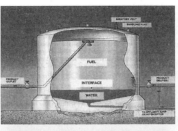

Figure 3. Storage tank microbial contamination resulting in growth proliferation, corrosion products and deterioration of fuel quality

The notable increase in supplies of microbially contaminated fuel are directly related to poor practices and housekeeping procedures.

Controls: relaxed controls and housekeeping standards at refineries, tank farms and delivery barges can be responsible for fuel contamination. Microbes may also contaminate storage tanks if these are washed out using polluted harbour and estuary waters.

Process: factors also contributing to microbial problems are those procedures implemented to achieve improved fuel production economy.

Production: the size, saturation and configuration of hydrocarbon molecules.

Product: fuel additives frequently contain nitrogen and phosphorus.

Distribution: faster throughput allows less time for particulates and water to settle.

The higher incidence of contamination in fuel is partly due to the predominant molecular types. Black distillates are more vulnerable due to the abundance of trace elements emanating from quantities of residual fuel and the cutter-stock used during blending.

Storage: viable microbial concentrations at the fuel/water interface will result in growth proliferation. The denser particulates will in time, fall out of suspension and accumulate on the storage tank floor. Storage tanks which are infected, as shown in Figure 3, will not always deliver 'unfit for use' contaminated fuel. Fuel drawn from the top of an infected but undisturbed tank will be reasonably clean. If the draw-off is taken by floating suction during a fast delivery when the tank level is low, this will result in heavily contaminated fuel. The outcome could be severe operational problems within a few hours of using the infected fuel.

2.3.3 Onboard

Figure 4. Fuel oil microbial spoilage by stages

There will always be an initial source of contamination to shipboard fuel, probably from delivered bunkers. Once infestation has occurred, particular circumstances may encourage microbial proliferation onboard.

Further inoculation is then immaterial as the key aggravating factors to incubate microbes are water and warmth resulting in microbial spoilage, as shown in Figure 4. The result is that a small number of microbial cells can multiply to produce a few kilograms of biomass in a short time.

2.3.4 Cargoes

There have been a number of incidents of severe microbial contamination of gas oil and kerosene cargoes, and one is described in Section 6, Case History 2. These incidents are usually the subject of litigation which tries to identify the source of the contamination and allocate culpability. These sources could be one or more of:

- The load port contamination of the storage tanks.
- The load port delivery pipe-line.
- The cargo tanks of the vessel.
- The pump room of the vessel.
- Sea water ingress.

An expert microbiological examination and evaluation of retained samples is necessary to investigate these incidents.

2.4 Symptoms of Microbial Contamination

Medium	Fuel
Visual	Aggregation of microbes into a biomass, observed as discolouration, turbidity and fouling Biosurfactants produced by bacteria promote stable water hazes and encourage particulate dispersion Purifiers and coalescers which rely on a clean/fuel water interface, may malfunction Tank pitting
Operational	Bacterial polymers may completely plug filters and orifices within a few hours Filters, pumps and injectors will foul and fail Non uniform fuel flow and variations in combustion may accelerate piston rings and cylinder liner wear rates and affect camshaft torque

Table1. Symptoms of microbial contamination of fuel

Visual and operational problems are commonly encountered after severe microbially contaminated fuel is used with all or some of the phenomena as listed in Table 1. The symptoms of microbial contamination, will confront the ship operator within a few hours. The extent of the contamination must first be established via the following procedures.

Samples from service, header and storage tank drains should be taken using clear glass bottles.

Contamination will be apparent as a haze in the fuel from the presence of sludge as shown in Figure 5. This sludge readily disperses in the fuel when the sample is swirled and sticky 'cling film' flakes may be seen adhering to the wall of the bottle. The water will be turbid and there may be some bottom sludge; if this is black, there are SRB present and they will be an added corrosion hazard.

Samples and the supplier's retained sample should be forwarded for microbiological examination to identify if the bunkers supplied were contaminated.

The ship operator can now evaluate the results and instigate an emergency strategy by using the cleanest fuel. If only heavily contaminated fuel is available, it should first be allowed to settle for as long as possible and then drawn off from the top to a clean tank, preferably via a filter, purifier, centrifuge or coalescer. Using a biocide at this stage is usually not advisable due to a tendency to block filters with the dead, dislodged biofilms.

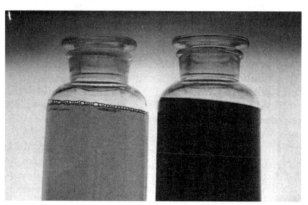

Figure 5. Contaminated fuel oil indicated by its haziness appearance

3. Effects of Microbial Contamination

Microbial contamination can produce a varied range of effects which can be directly attributable to microbial spoilage and corrosion.

3.1 Composition of Fuels

Increased microbiological contamination of distillate fuels such as gas oil, diesel and kerosenes (rather than residual fuels) is largely due to changes in the fuel chemistry and the widespread use of fuel additives. Fuels suffer from microbiological attack (oxidation) due to their chemical composition. They contain a wide range of compounds (diesel contains more than 250) but are primarily composed of n-alkanes (50% mostly with carbon numbers between 10 and 18), other straight chain, branched and cyclic alkanes, and aromatic hydrocarbons (eg benzene, toluene, xylene) in variable proportions of up to 25%, and polycyclic compounds.

The fuel, therefore, has abundant carbon sources for microbiological growth but is deficient in inorganic nutrients such as nitrogen, phosphorous and potassium. These elements are often the limiting factors in microbial degradation of the fuel and must be supplied from an external source, such as fuel additives and polluted harbour and estuary waters.

At every bunkering, the microbial food source is replenished but there is rarely sufficient contact time for significant chemical changes in the fuel due to microbes.

3.2   The Microbial Spoilage Process

Figure 6. Fuel oil filter blockage from clean, initial and severely contaminated by stages

Until recently, fuels seemed better able to accommodate minor degrees of microbial contamination without ship operators experiencing functional difficulties.

Evolution of microbe species has spawned new types of bacteria in fuels, which produces sticky polysaccharide polymers, similar to 'cling film'. This rapidly clogs filters by trapping particulate matter, as shown in Figure 6.

Severe and prolonged microbial activity will result in the fuel being degraded, reducing the hydrocarbon chain length and thereby reducing the fuel's overall calorific value. Furthermore, microbial metabolites, such as hydrogen sulphide gas, can cause 'souring' of the fuel. When altering the fuel's hydrocarbon molecular structure, both chemical and physical changes are observed affecting the pour point, cloud point and thermal stability. In addition, bacteria which are prolific producers of biosurfactants will, after time, establish a stable water haze. The fuel may eventually fail specification tests for water separation.

These fuel conditions will result in operational problems.

3.3　The Microbial Corrosion Process

Steel perforation by pitting corrosion, particularly in tank bottom plates, is now routinely observed where microbes accelerate the usual chemical and electro-chemical corrosion mechanisms. Where infection is severe, corrosion from organic acids produced by moulds may occur, particularly on aluminium alloys. These mechanisms have been described in Chapter 1.

4.　Prevention and Elimination of Microbial Contamination

A number of factors control the rate at which microbiological problems will develop and indeed, whether problems will develop at all. Without infestation of a system, microbial problems cannot arise. Unfortunately, this is never the case and low numbers of viable microbes will always find their way into a system. If they reproduce slowly, the progeny will be removed whenever water is removed and large numbers of microbes will not accumulate. However, a large influx of microbes resulting in a heavily contaminated system will defeat this simple self-regulating mechanism; there are more microbes to reproduce and the potential for future problems will have been established.

4.1　Prevention

Prevention of microbial infection may be controlled by good shipboard practices supported by physical cleaning and chemical control treatments.

4.1.1　Physical Prevention

Water is critical for microbial growth and its presence always presages the potential for contamination. Certain procedures are recommended to avoid microbiological problems:

- Assess and modify inadequate fuel tank water bottom drainage systems.
- Implement a regular and effective fuel tank water drainage programme.
- Prevent long term fuel storage and stagnation by alternating fuel tanks.
- Periodically clean and chemically disinfect fuel systems, purifiers, filters and coalescers.
- Isolate service fuel tanks against contamination of suspect infected bunkered fuels.
- Routinely inspect fuel tanks for microbial biofilm slimes and paint and corrosion damage.

- Procure suitable fuel test kits to evaluate microbial levels.
- Monitor fuel suppliers' quality and consistency of bunkers.

However, even the most meticulous maintenance cannot completely eliminate the potential for microbial problems.

4.1.2   Chemical Prevention

Once a ship's system has become heavily contaminated, the infecting microbes will pose an ongoing potential hazard and final resolution of the problem will involve the use of anti-microbial chemical biocides. Preservative chemicals protect against possible minor infections over a prolonged period, being slow-acting but persistent. They will not necessarily cope with major contamination. Any chemical used as a preservative to prevent the proliferation of organisms must be sufficiently soluble to migrate into, and disperse within, the designated fuel and/or water phase. These chemicals should be amenable to simple concentration monitoring as intermittent use could encourage the emergence of resistant microbes, exacerbating existing problems.

Fuel preservatives: fuel-soluble chemicals are applied directly via the tank top to disperse throughout the fuel. Sufficient chemical must be dissolved in the fuel to infiltrate inaccessible structures within the fuel tank and must persist if long term preservation of a fuel is required,.

Water preservatives: water-soluble chemicals can be introduced by way of the tank top in water-soluble sachets or injected into the tank bottom. If long term protection of a tank is required, sufficient chemical must remain in the water bottom despite many refuellings.

4.2   Elimination

Elimination of established microbial infection may be accomplished by physical methods and/or chemical dosing treatment.

4.2.1   Physical Decontamination

Microbes do not die naturally and they must be killed or removed. Certain procedures are recommended to ameliorate existing microbiological problems:

Settling: microbes are denser than fuels and will gravitate towards the tank bottom. Fuel cleanliness is dependent upon the size and degree of aggregation of the microbes and the time for settlement, as shown in Figure 7.

Centrifuges: microbes subjected to centrifugal forces will separate out. Fuel treatment is effective and is dependent upon the density and degree of aggregation, fuel viscosity and the retention time, as shown in Figure 8.

Filtration: microbes with cell sizes of a only few μm, may be removed by filters. Successful fuel filtration is dependent upon using a series of appropriate filters and even if the pore sizes are greater than the microbe size, microbial aggregates can be removed.

Heat: microbes exposed to heat in excess of 70°C and for 20 seconds will kill the majority of spoilage microbes. Fuel sterilisation is permanent and the rate is dependent upon temperature and time of application.

Figure 7. Contaminated fuel oil after being allowed to settle

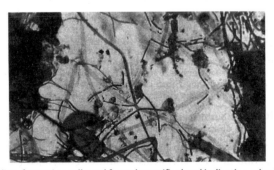

Figure 8. Slime formation collected from the purifier bowl indicating mixed microbes

4.2.2 Chemical Decontamination

The requirements for an effective biocide which will kill organisms, clean surfaces and suppress re-infection are given below.

Fuel biocides: these biocides have a finite life, being expendable either due to absorption by the microbes they kill, or by leaching from the fuel into the water phase. They should only be used if they are chemically compatible with the fuel, the machinery and system components.

Water biocides: they are applied to destroy a wide range of microbiological species, acting efficiently on the tank bottom area as well as at the fuel/water interface. They may contain dispersing or sludge-solubilising agents to remove deposit accumulation.

Moderately contaminated systems: intermittent dosing with a fuel biocide will prevent serious fouling or malfunction and enable the fuel to be used during and after decontamination. After the addition of the biocide, the fuel should be circulated through all parts of the system.

Severely contaminated systems: shock dosing with a high concentration of fuel biocide is essential. This will result in detachment of biofilms which settle as sediment on the tank bottom. This deposit must be removed by physical decontamination methods to prevent blocked fuel systems when the fuel is used. Physical cleaning of the empty tank will be necessary to remove all traces of microbial aggregates.

5.   Health of Crews

5.1   Microbial Hazards

Microbes are included as 'substances hazardous to health' in the UK 'Control of Substances Hazardous to Health Regulations' (COSHH), 1994. It would be prudent to comply with these industry regulations.

The bacterium *Pseudomonas aeruginosa* and the mould *Aspergillus fumigatus* are sometimes detected in fuel. These are common spoilage microbes with a potential for causing clinical infections in susceptible individuals. SRB in fuel tank water bottoms after extended storage and stagnation will produce pungent and toxic hydrogen sulphide. Dangerous and lethal levels of hydrogen sulphide within enclosed spaces should be avoided. Ventilation is certainly not a guarantee of safety, as SRB will continue to generate hydrogen sulphide until they are killed - thus entry into a contaminated region is not advisable.

5.2   Chemical Biocides

Biocides are designed to kill or prevent microbial infestation and are toxic chemicals. The application of these chemicals should be controlled and in accordance with the health and safety material data sheets supplied by the manufacturer. A risk assessment will be necessary and a 'duty of care' will exist towards any third party handling fuel and sludge containing biocides

5.3   Test Kit Disposal

Test kits comprising nutritive substances are designed to promote microbial growth and to enhance their visible presence. Due diligence should be taken to avoid direct contact with the cultured microbes and disposal of used kits should be in accordance with the manufacturer's instructions.

6. Cost to Industry

6.1 Marine Casualties of Microbial Contamination

Case History No. 1.

Reported problem: Gas oil fuel contamination.
Microbial presence: Bacteria, yeasts, moulds and SRB.
Vessel: Patrol ship.
Built: Denmark - 1960.
Investigation by Lloyd's Register of Shipping.

The vessel loaded gas oil bunkers from a North Atlantic port and, in accordance with procedures, used one of the settling tanks to store the fuel until a sample was analysed as being 'fit for use'. Analysis of the sample indicated that the gas oil was unsuitable for service due to severe microbial infection. The microbial count for bacteria, yeasts, moulds and SRB was greater than 10 000, in the region of 10^6, in a litre sample.

Evaluation

Had the vessel used the fuel, the spoilage material would have caused filter blockage. The vessel duly offloaded the fuel and implemented a cleaning and biocide treatment programme. In this instance, all costs were against the supplier.

Case History No. 2.

Reported problem: Gas oil cargo contamination.
Microbial presence: Bacteria, yeasts and moulds.
Vessel: Product tanker.
Built: Sweden - 1975.
Investigation by Lloyd's Register of Shipping

The vessel had loaded a microbially-contaminated gas oil cargo of 13 000 tonnes from a Baltic port. Routine testing of vessel's cargo before discharge in a European port confirmed the presence of severe microbial infection. After the cargo was declared spoilt, it was subsequently discharged. Since the discharge, the vessel had implemented tank-cleaning procedures with chemical dosing to eliminate the continued presence of microbial contamination but without success. Cargo samples received from the vessel enabled the types of microbes to be identified and the choice of biocide made. The microbial count for bacteria, yeasts and moulds was greater than 25 000 in a litre sample. Onboard assessment of the operating problems, tank layout, piping systems and the available stand-down period, allowed a tank decontamination and a time schedule programme to be planned and implemented. Cold water, dosed with the correct biocide concentration and circulated using the high pressure Butterworth system, at 1m, 4m and 8m from the tank tops, dislodged the adherent biomass films from the tank walls. Further soaking of each tank ensured that the biocide kill time of six hours was achieved, such that the entire operation was completed in 48 hours, as shown in Figures 9 and 10.

Evaluation

Application of the correct biocide and strategies to eliminate the persistent microbial infection proved effective and a 'microbial-free cargo tank' report was awarded to the vessel. Until such a report was issued, the vessel was not acceptable to charterers. Direct and indirect costs due to loading a contaminated cargo and decontaminating the vessel were in the order of £70 000, which exceeded the profit in carrying the tainted cargo.

Figure 9. Viable microbial biofilms adhering to cargo tanks before biocide decontamination

Figure 10. Dead microbial biofilms floating on water after biocide decontamination

7.   Identifying Microbial Presence

Routine sampling and microbiological tests should be carried out using onboard tests and supported, if necessary, by a comprehensive laboratory report. The objective of microbiological testing is to detect and identify if microbes (bacteria, yeasts, moulds and SRB) are present and viable, then to monitor subsequent anti-microbial strategies. Section 6, Chapter 1, describes the laboratory procedures which are available for testing fuel, lubricants, etc and any associated water phase, and also describes onboard test procedures in detail. Those tests suitable for investigating fuel and fuel systems have been identified in this section.

There are numerous types of dip-slide test kits available which can be used for the analysis of aerobic microbes in the water phase and the fuel phase of bunkers as well as for yeasts and moulds. For SRB the use of a gel test kit will be required. These results will provide initial evidence that the fuel is/is not 'fit for use'. It is usually prudent to test water phase (but not fuel phase) for anaerobic SRB. If a sample contains both fuel and water, these should be tested separately.

Laboratory analysis of the sealed retained bunker sample will provide supportive evidence and, should operational problems be encountered onboard, laboratory documented evidence for consequent claims against the supplier.

7.1   Sampling Procedures

The correct procedure, container and sampling location is important as severely infected delivered fuel will result in immediate operational problems. The most heavily infected fuel will be found in tank bottom and drain samples. Proliferation of moderate microbial infection from low numbers to numbers likely to cause problems is not rapid, usually taking several months provided that free water is eliminated.

Routine sampling of delivered fuel

Prior to delivery, bottom drain water samples from the storage tank, road tanker and barge will enable a rapid analysis test to be undertaken. If no water is recovered, a bottom fuel sample is most likely to reveal contamination, although this fuel sample will not be suitable for rapid analysis. During delivery, sample from the loading pipeline, hose or manifold at the beginning and end of bunkering.

Routine sampling of onboard fuel

Engine room tanks should be sampled at three-monthly intervals. Accumulation of water in engine room tanks and observed filter blocking problems, requires more frequent sampling. Main storage tanks need only be sampled every six months and auxiliary tanks and lifeboat tanks sampled annually.

Sampling procedures and sample handling are referenced and published by the Institute of Petroleum as guideline methods.

7.2   Sample Testing Onboard

7.2.1 Tests of the Water Phase in Bunkers

Test for bacteria: dip-slides for detecting bacteria usually incorporate a dye which stains the bacterial colonies red. After incubation in a warm environment for two to three days, the result is read from a calibration chart. If there is insufficient water to 'dip' the slide, a sterile disposable pipette can be used to apply water over the slide. Sea water always contains microbes and the result must be interpreted with typical numbers per millilitre in mind:

Clean sea water	10^2-10^3
Polluted water	10^3-10^4
Moderately infected fuel in water bottom	10^5
Severely infected fuel in water bottom	10^6-10^8

Test for moulds and yeasts: dip-slides usually have one slide surface for culturing bacteria and the opposite side for yeasts/moulds. On the side for culturing yeasts and moulds these microbes grow as round or furry colonies. After incubation in a warm environment for three to five days, the result is read from the calibration chart and typical numbers per millilitre are:

Clean sea water	0 -10^2
Polluted water	0 -10^3
Moderately infected fuel in water bottom	10^3-10^4
Severely infected fuel in water bottom	10^4-10^6

Tests on surfaces: dip-slides can be pressed against tank surfaces and then incubated to obtain an indication of cleanliness. This is a useful strategy to check the efficacy of a decontamination procedure and typical numbers are:

Clean/moderate surface condition	10 -10^2
Moderate/dirty surface condition	10^2-10^4
Dirty surface condition	10^4-10^6

Tests for SRB: dip-slide-type tests are inadequate as these microbes grow only in the absence of oxygen and require an aqueous sample to be inserted into a gel. There are numerous types of gel test kits available which can be used for the analysis of anaerobic SRB. The degree of infection is determined by the rate of development and intensity of black colouration. After incubation in a warm environment for up to 10 days, any positive result should be viewed with concern.

7.2.2 Tests of the Fuel Phase of Bunkers

Onboard dip-slides are designed for tests of the water phase, and the chart results are calibrated for water samples. They should not be used for fuel phase samples. Commercially-available tests kits for the fuel phase are available, which will identify microbial presence within a light, moderate and heavy contaminated fuel range.

Laboratory sample testing methods are published by the Institute of Petroleum in the procedures IP 385/99 and IP 472/02.

8 Standards

There are no regulatory microbiological standards which can be applied to microbially infected bunkered fuel, which renders the fuel 'unfit for use' and which can have a significant impact upon the safe operation of ships. The Institute of Petroleum's Guidelines for the Investigation of the Microbial Content of Fuel recommends that there should be no fixed standards but that acceptable numbers should be related to the sampling point and the purpose for use. For example, tank bottom samples will always be more contaminated than top samples, and filtered fuel delivered to the engine should be cleaner than fuel in double bottom tanks.

8.1 Perception of Fuel Quality

There is no doubt that the marine industry continues to lag behind other major industries in its appreciation of the consequences of microbial infection. The experience gleaned from the detection, quantification and cure of infection in other industries can still be applied in principle, but there are a variety of additional factors unique to the marine industry. Resolution of fuel contamination problems is not addressed in the present ISO 8217 1996 fuel standards, since these do not specify microbial levels, stating only that the quality of fuel should be 'fit to use'.

8.2 Fuel Supplier's Responsibility

The notable incidences of microbially-infected bunkered fuel supplied to ships, does little to improve the confidence of ship operators in the marine industry. In every line of business it is natural to seek to maximise profit and competition may tempt suppliers to cut out what they view as 'unnecessary' expenditure. Allied to this is the inescapable fact that the running costs of a ship are high and returns are limited, thus encouraging fierce competition and creative interpretation of responsibility, usually in ignorance of the consequences to the safe operation of the ship and the crew. The ISO 8217 fuel standard is an adequate indication of merchantable quality. That the standard does not list every attribute required of a merchantable fuel does not mean that the supplier is free to deliver unmerchantable fuel. Nor does it mean that the fuel complying with all tests listed but which for other reasons is unsuitable, should be regarded as merchantable. After all, the permissible content of radioactive material is not listed in any fuel specification, which is not a carte-blanche for its inclusion, as it is obvious that its presence is most undesirable. Similar logic must be applied to microbial contamination as being unacceptable and although not specifically mentioned in the current standards, contaminated fuel should not be supplied to ships.

Since fuel suppliers are not party to what ship operators intend to do with the loaded fuel, be it for immediate use or long term storage, they cannot be the ones to decide microbial limits. By definition, microbial contamination should not result in engine damage and treatment plant operational problems. The marine industry should not be treated on the mentality of 'out of sight, out of mind', leaving the false impression that ship operators are responsible for resolving supply problems. More attention should be directed towards suppliers' poor 'housekeeping' practices which is where many microbial contamination problems begin.

8.3   Ship Operator's Responsibility

The best remedy at present would be for ship operators to persuade suppliers to provide 'microbe-free' clean fuel. The onus is then on the ship operator to ensure that microbial problems do not establish themselves onboard. There is also more awareness by fuel purchasers, who now recognise that agreement contracts require a clause stating that 'products are to be microbe free'. Although not precise in its definition, its presence should reduce the numerous cases of contaminated supplies. The ethos that 'prevention is better than cure' should be adhered to, and fuel suppliers selected for the quality and consistency of their products.

9   Conclusion

Microbial fouling and corrosion are becoming commonplace onboard and will probably continue to increase due to fuel oil formulation and handling trends, polluted harbour waters plus regulatory pressures. Ship design and construction should recognise these facts, so that systems are designed to be less conducive to harbouring microbes. Onboard test kits are available for detecting and quantifying microbial contamination and for monitoring the success of anti-microbial measures. Various strategies are available for reducing the problems, but solutions must be safe and environmentally acceptable. The details needed to build these strategies into full working protocols will depend upon individual circumstances, the time available and access to anti-microbial agents. There are many different microbiological problems which need to be met with specific, tailored solutions. However, there are certain common principles of good practice which can be implemented. These anti-microbial strategies are detailed in this chapter and have proved to be successful in the field.

Briefly recapping, these are:

- Physical prevention: preventing ingress of inoculating microbes, particularly those already adapted to growth in relevant environments. Avoid spreading contamination by passing clean fluids through contaminated pipes, filters and into dirty tanks. Minimise conditions which encourage water accumulation and microbial growth.

- Physical decontamination: settling, heat, filtration and centrifugal procedures all aid in decontamination, the choice depending upon equipment and time constraints.

- Chemical prevention: protecting against minor contamination, coupled with good housekeeping to prevent rather than cure infection.

- Chemical decontamination: a wide range of chemical biocides are available. Only a few are appropriate to each specific application, there being no universal eradication fluid. All are toxic and must only be used with due regard to health and safety and environmental impact.

Microbial investigations by Lloyd's Register of Shipping have identified that we are no longer just looking at dirty fuel oil and blocked filters etc, but that microbial contamination is now compromising the safe operation of the ship.

10. Acknowledgements

The author would like to thank the members of The Institute of Marine Engineers Microbiological Technical Sub-Committee, The Institute of Petroleum Microbiological Fuels Group Committee and Institute of Corrosion Microbiological Corrosion Unit Committee for their help in providing input to the chapter and to Dr S Harold and J Heath for reviewing the chapter. The author also thanks Lloyd's Register of Shipping for its permission to publish this chapter taken from the LRTA paper *Microbial Attack on Ships and their Equipment* paper No.4, 1994-95. However, all views expressed are those of the author.

The author would like to acknowledge more fully the assistance provided by Echa Microbiology Limited, Oil Plus Limited, Rohm and Hass (UK) Limited, Boots Chemicals Limited, contributors and other members of the Institute of Marine Engineers Microbiological Technical Sub-Committee. This chapter could not have been produced without their contributions. The specific figures and quotations provided are listed below:

Echa Microbiology Limited

The table provided by Echa Microbiology Limited to this chapter is Table 1.

In addition, the chapter quoted parts from a number of Echa-related publications as follows:

Microbiological Problems in Distillate Fuels, Trans. I.Mar.E., 104, 1992.

Microbial Proliferation in Bilges and its Relation to Pitting Corrosion of Hull Plate of Inshore Vessels, Trans. I.Mar.E., 105(4), 1993.

Microbes in Fuels, Lube Oils and Bilges; Recognition and Monitoring, Seminar Workshop, I.Mar.E, 1993.

Safe Acceptable Anti-Microbial Strategies for Distillate Fuels, Conference on Stability and Handling of Liquid Fuels, 1994.

Bugs in Fuels - Time for Action, MER, 1996.

Oil Plus Limited.

The figures provided by Oil Plus Limited to this chapter are; Figures 2, 6 and 8.

In addition the chapter quoted parts from a number of Oil Plus-related publications as follows:

Microbial Problems in the Offshore Industry, Oil Plus reference 110147, 1996.

Assessment, Monitoring and Control of Microbiological Corrosion Hazards in Offshore Oil Production Systems, NACE, 1987.

Monitoring and Control of Microbes, Cost Effective Ways to Reduce Microbial Corrosion, Elsevier Applied Science, 1988.

Rohm and Hass (UK) Limited.

The figures provided by Rohm and Hass (UK) Limited to this chapter are; Figures 1, 3 4, 5 and 7.

In addition the chapter quoted parts from a number of Rohm and Hass publications as follows:

Fuel Protection Microbicide, Technical Bulletin, 1992.

Hydrocarbon misadventures, Technical Bulletin, 1995.

Boots Chemicals Limited.

The chapter quoted parts from a number of Boots publications as follows:

Microbial Control in Exploration and Production, Technical Bulletin, 1995.

Laboratory and Field Experience with a Biocide in Industrial Water Systems, Chemspec, 1986.

11. References & Bibliography

1. Stuart, RA, *Microbial Attack on Ships and their Equipment* LRTA paper No.4, 1994-95.
2. Stuart, RA, *Safety of Ships, Health of Crews and Cost to Industry*, I.Mar.E Seminar Workshop, 1996.
3. Hill, EC and Hill, GC, *Microbiological Problems in Distillate Fuels*, Trans. I.MAR.E., 1992.
4. Hill, EC and Hill, GC, *Microbial Proliferation in Bilges and its Relation to Pitting Corrosion of Hull Plate of In-Shore Vessels*, Trans. I. MAR.E., 1993.
5. Hill, EC and Hill, GC, *Microbes in Fuel, Lube Oils and Bilges; Recognition and Monitoring*, I.Mar.E Seminar Workshop, 1993.
6. Hill, EC, *Safe Acceptable Anti-Microbial Strategies for Distillate Fuels*, 5th International Conference on Stability and Handling of Liquid Fuels, Rotterdam, 1994.
7. Hill, EC, *Bugs in Fuels - Time for Action*, MER, 1996.
8. Sanders, PF and Stott, JFD, *Assessment, Monitoring and Control of Microbiological Corrosion Hazards in Offshore Oil Production Systems*, NACE, Corrosion Seminar 87, Paper No. 367, 1987.
9. Sanders, PF, *Monitoring and Control of Microbes, Cost Effective Ways to Reduce Microbial Corrosion*, Publication Elsevier Applied Science, 1988.
10. Smith, RN, *Microbiology of Fuels*, I.P. Microbiology Conference, 1986.
11. Smith, RN, *Bacterial Extra-Cellular Polymers; a Major cause of Spoilage in Middle Distillate Fuels*, I.P. paper, 1988.
12. Guthrie, WG, Elsmore, R and Parr, JA, *Laboratory and Field Experience with a Biocide in Industrial Water Systems*, Chemspec 86, 1986.
13. Chesnau, HL and Doris, MM, *Distillate Fuel; Contamination, Storage and Handling*, ASTM STP 1005, 1987.
14. Shennan, JI, *Microbiological Risk Assessments for COSHH*, I.P. Microbiology Conference, 1990.
15. Cordruwisch, R, Kleinitz, W, Widdel, F, *Sulphate Reducing Bacteria and their Economic Activities*, Society of Petroleum Engineers - USA, 1985.
16. Lloyd's Register, *Rules and Regulations for the Classification of Ships*, 1998.
17. International Maritime Organization, *Regulations for the Prevention of Pollution by Oil*, MARPOL 73/78 Annex 1, 1998.
18. Health and Safety Executive, *Entry into Confined Spaces*, HMSO 1977 ISBN 0 11 883067 8.
19. Health and Safety Executive, *Respiratory Protective Equipment; a Practical Guide for Users*, HMSO 1990 ISBN 0 11 885522 0.
20. Health and Safety Commission, *Safe use of Pesticides for Non-Agricultural Purposes, Approved Code of Practice*, HMSO 1991 ISBN 011 885673 1.
21. Health and Safety Executive, *Control of Substances Hazardous to Health:*

Control of Carcinogenic Substances; Control of Substances Hazardous to Health Regulation (COSHH) 1994, Approved Code of Practive, HMSO.

22. Institute of Petroleum, *Determination of the viable Microbial Content of Fuels and Fuel Components Boiling Below 390 deg.C - Filtration and Culture Method IP 385/99 and Determination of Fungal Fragment Content of Fuels Boiling Below 390 deg.C IP 472/02.*

23. Institute of Petroleum, *Guidelines for the Investigation of the Microbial Content of Fuel Boiling Below 390 deg.C and Associated Water.* Sampling procedures and sample handling, 1995.

3. Lubricant and Hydraulic Oils

EC Hill

Contents

1.   The Microbial Problem

In this section the problems which are discussed occur in oil/water mixtures in which water is the minor but essential component (usually about 0.1%), and is present as a contaminant. The oil phase is nutritionally rich for microbes and capable of sustaining very large populations. The oil may contain about 20% of additives, some of them incorporating the vital nutrient elements nitrogen and phosphorus. More nutrients may be available in the contaminating water, particularly if it has been treated with anti-corrosive agents. Although the infection may initially be restricted to the oil/water interface in the tank or sump bottom, infected water droplets and microbes eventually become dispersed throughout the oil phase. The problem takes the form of fouling, spoilage and corrosion. Ships' engine crankcase lubricants have been particularly prone to infection[3,4] as they are inevitably water-contaminated and maintained at temperatures conducive to microbial growth, about 38-50°C. Steam turbine lubricating oils are also prone to water contamination and hence are susceptible to infection: microbial problems have also been experienced in a variety of other 'straight' oils such as cutting oils, bearing oils, hydraulic oils and rust preventive oils. The phenomenon was rarely apparent before the mid 1970s partly due to the simpler composition (and lower nutritive value) of oils up to that time, but other factors were identified in ships' crankcase oils which could explain the sudden onset of microbial problems. In most cases, engine cooling water was the prime source of water which the microbes need, and changes have occurred in coolant additives which are also significant. In total, important factors were:

- Introduction of sophisticated, nutritious and biologically 'soft' oil formulations.
- Increased use of non-toxic corrosion inhibitors in engine coolants; chromates used previously were actively anti-microbial and prevented growth when cooling water leaked into the crankcase oil.
- Changes in base oils which favoured microbial growth.
- Reduced use of heated renovating tanks which had previously acted as intermittent oil sterilisers.
- Ineffective purification.
- Neglect during lay-up.

The oil user can experience substantial system fouling, changes in the physical and chemical characteristics of the oil, corrosion and equipment malfunction, sometimes of a catastrophic nature. Additive degradation removes the beneficial properties which these would normally impart; alkalinity (Total Base Number) decreases; oil viscosity may change; microbial surfactants promote stable dispersions of water droplets into the oil. In extreme infections of crankcase oil, bearings may fail and the engine seizes[3]. Some of these effects are illustrated in Figures 1, 2 and 3. All or some of the following phenomena would suggest that significant microbial growth was occurring in a crankcase oil:

- Unusual smells.
- Slimy oil on dip-stick, crankcase sides or in a sample.
- Excessive sludge in sump bottom or at purifier discharge.
- Stable water haze in oil (due to microbial surfactants).
- Honey coloured film and/or corrosion pitting on journals.
- Black stains on white metal.
- Unusual rusting.
- Corrosion/erosion of purifier bowl.
- Fall in oil TBN or onset of acidity.
- Increased wear rates.
- Filter plugging.
- Paint softening.

Many of these phenomena could occur in turbine or bearing oils. These symptoms could have a non-microbiological cause, and confirmation of a micro-biological problem by sampling and testing is advisable.

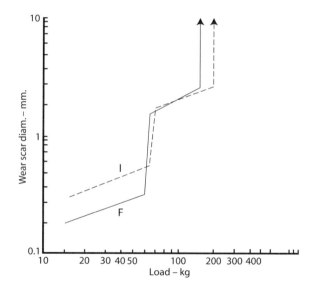

Figure 1. Four-Ball Wear (IP 239/73) test results on fresh oil (F), and oil in which a microbe (Aspergillus) has been grown in the laboratory for four weeks at 30°C (I). Comparing I to F wear scars are double the size when surface contacts are lightly loaded and there is a slight increase in the load which can be applied before complete lubrication failure occurs. (After Hill[4]).

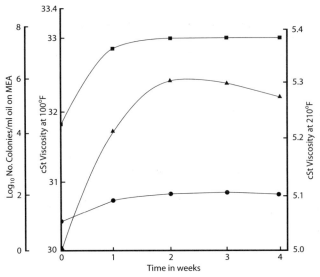

Figure 2. Viscosity changes when microbial growth occurs in a base oil held at 30°C in the laboratory. (After Hill and Al-Haidary[2]). ● Viscosity (centistokes) at 100°F. ■ Viscosity (centistokes) at 210°F. ▲ Log of numbers ml⁻¹ of yeast (Candida)

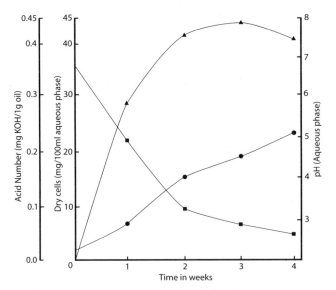

Figure 3. Acidity produced by growth of Candida in base oil/water held at 30°C for four weeks. (After Hill & Al-Haidary[2]). ■ pH of aqueous phase. ● Acid Number (by potentiometric titration) of oil expressed as mg KOH g oil. ▲ Dry weight of Candida measured in the aqueous phase and expressed as mg 100ml⁻¹

Hydraulic oils, to function reliably, must be free of particles and hence are very sensitive to microbial fouling which will cause equipment malfunction. Infection in complex hydraulic systems can be very localised.

1.1 Causative Microbes

'Straight' oils tend to be alkaline with a reserve of alkalinity and are either maintained at an elevated temperature or become hot during use. These conditions favour the selection of bacteria of a type termed Gram-negative bacteria, such as *Pseudomonas* spp, which are the predominant spoilage agents in crankcase oils. Occasionally thermotolerant moulds, particularly *Aspergillus* spp, cause problems. Yeasts are frequently dominant in hydraulic and bearing oils.

Engine lubricants in use are usually agitated and well-oxygenated; hydraulic oils are less agitated but contain oxygen dissolved under pressure. The microbial population will change when systems cool and become stagnant and oxygen deficient. Sulphate reducing bacteria may then flourish and generate corrosive sulphide.

1.2 Sources of Microbes

This is often obvious, for example by filling new oil into dirty systems or passing it through dirty pipes or filters. Frequently the microbes are already flourishing in the water phase which leaks into the oil, particularly engine cooling water and sea water. Occasionally the spare oil tank becomes contaminated and this spreads to the oil charge in use.

2. Monitoring and Investigating Microbial Problems

The symptoms of microbial contamination listed in section 1 will only be apparent when very substantial microbial growth has occurred. In a very large oil system, several tonnes of microbial sludge may be present. There are ways of curing these severe problems and they are addressed in sections 3.1.3 and 3.1.4. It is advisable to determine the extent and severity of the infection by taking and testing a range of samples so that the best decontamination strategies can be implemented. However, sampling and testing is of most value if it is used to detect the earliest stages of an infection so that remedial measures can be taken in good time and operational problems avoided.

2.1 Sampling

Most oil systems run at levels of microbial infection which are negligible or can be tolerated. For routine checks on contamination levels it is sufficient to test the sump bottom (where growth usually starts) and unfiltered circulation oil, to see how much of any growth is being dispersed. These results may be accepted as

Oil from Sump Bottom or Before Purifier					Circulation Oil Before Filter			Possible Action
MicrobMonitor², Total microbes ml⁻¹			Sig Sulphide test		MicrobMonitor², Total microbes ml⁻¹			
$\langle 10^3$	10^3-10^4	$\rangle 10^4$	No SRB	SRB*	Negative ($\langle 10^2$)	10^2 -- $\langle 10^4$	10^4 or more	
✓			✓		✓			No action. Re-test in 1-3 months.
	✓		✓		✓			Check coolant for contamination. Decontaminate if necessary. Re-test oil in one month.
		✓	✓		✓			Decontaminate sump bottom with biocide via sounding pipe. Check coolant and decontaminate if necessary. Re-test oil after treatment.
✓ or	✓			✓	✓			
✓ or	✓		✓ or	✓		✓		Take oil samples before and after purifier and test. Refer to Table 2. Decontaminate whole system when convenient. Check coolant and decontaminate if necessary. Re-test oil after treatment.
✓ or	✓		✓ or	✓			✓	Test oil samples before and after purifier. Refer to Table 2. Decontaminate system ASAP. Check coolant and decontaminate if necessary. Inspect for corrosion. Re-test after treatment.
✓ or	✓			✓		✓ or	✓	

* Any positive for SRB increases the urgency of action.

Table 1. Tests on crankcase oil in circulation and in sump bottom

normal or must be interpreted and acted upon, for example by further sampling and testing, by local treatment with biocide or heat, or by anticipating action but postponing it until the next sampling time to establish whether there is an upward trend. An opinion on the interpretation and possible actions for routine test results is given in Table 1. If any anti-microbial action is required or antici-pated it is rarely possible to plan this logically without taking and testing sam-ples from several parts of a system. This is particularly important when systems incorporate a microbe limiting device such as a purifier or a filter. Crankcase oil systems and associated cooling systems are complex. A simplified layout is given in Figure 4 to illustrate that microbes will flourish in some wet and warm loca-tions but will be killed or removed in other locations. A complete set of samples

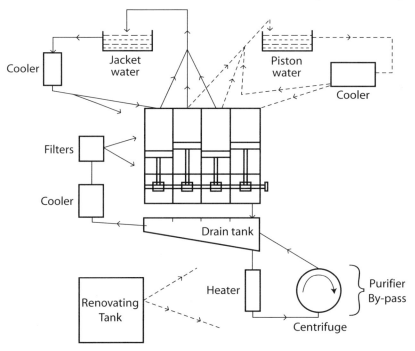

Figure 4. Simplified block diagram of cross-head diesel engine. Microbes could be killed or removed at the purifier, the renovating tank and the filter and could grow in other parts of the engine.

for the thorough investigation of a suspected problem in the crankcase oil of a large marine diesel engine could be: sump bottom, before and after purifier, spare oil charge, before and after filter, associated cooling water systems, renovating tank if in use.

The significance of the results and the planned actions must be related to the sampling locations; two critical results are those on samples taken before and after the purifier. The purifier has been designed to reduce the viscosity of the oil by heating it and then centrifuging it to remove water and particles. The purifier can additionally have two important anti-microbial functions. At the centrifuge, many microbes which are entrained in water are removed with the water; others, suspended in the oil, are removed as far as the viscosity and retention time permit (the microbes have a density of about 1.05 g/cm^3). The most impor-

MicrobMonitor^2, Total microbes ml^{-1}							Possible action at purifier
Before purifier heater				After purifier centrifuge			
Negative	10^2	10^3	10^4 plus	Negative	10^2	10^3 plus	
✓or	✓or	✓		✓			No action at purifier
			✓		✓		Confirm heater at 75°C plus. Purify continuously
	✓	✓ ✓	✓		✓ ✓	✓ ✓	Confirm heater 75°C plus. Check turnover is less than 10 hours. Check retention time is more than 30 seconds. The smaller the kill the more urgent and thorough the action taken.

Table 2. Tests on oil passing through the purifier

tant anti-microbial effect can be at the heater which should have a pasteurising effect. Both the contact time and temperature are important. In Table 2 some possible results from tests on oil before and after the purifier are given and interpreted. More information on the role of the purifier is given in section 3.1.1.

Sampling a hydraulic oil system is less complex than crankcase oil but at least the system oil, the header tank oil and a header tank drain sample should be tested and also, if possible, the deposit on a filter. Infections may be localised to oil associated with water ingress, for example when submerged lengths of sub-sea hydraulic systems are connected, and these may be difficult to sample and detect.

A range of anti-microbial measures, both physical and chemical, could be deployed (based on sampling and test results) and expert help might be necessary to devise and implement the most appropriate and safe strategy.

In section 3 some suggestions are made for resolving complex problems with crankcase oil; many of them can be extrapolated to other oil systems. First it is appropriate to consider what microbiological test methods, particularly onboard test methods, are available.

2.2   Testing

2.2.1   In the Laboratory

Whilst any free water phase can be tested directly with conventional microbiological procedures, including those for sulphate reducing bacteria, the oil phase must first be converted into an emulsion by mechanically mixing it with a sterile aqueous solution of a non-toxic, non-ionic emulsifier[1]. This emulsion is then tested by conventional microbiological methods. If a laboratory is contracted to carry out tests on oils they should be requested to follow this procedure or use MicrobMonitor[2] tests.

2.2.2   Onboard Tests

A comprehensive range of tests suitable for onboard use have been reviewed[10]. Those suitable for testing marine lubricants were identified by Hill in a shipping journal in 1996. However the latest advice should always be sought as new test kits and methods are being developed. The current range of test kits available and the state of the art at this time is contained in Chapter 1. It is doubtful if the cost of new rapid methods (which are instrument based) can be justified onboard, and four simple procedures for testing lubricants and water associated with lubricants are proposed below. They are all described in more detail in Chapter 1.

MicrobMonitor[2] test on oil samples
A standardised 10µl loopful of oil is transferred to a bottle of nutritive thixotropic gel using the sterile loop supplied. The bottle is shaken to liquefy the gel and hence disperse the sample. After 'incubating' the bottle for one to four days, each microbe present multiplies to form a visible purple 'colony' of growth. Colonies are counted and the number equates to the number of microbes originally present in 10µl of oil.

Dip-slide test on sump bottom water or cooling water
Aerobic bacteria, yeasts and moulds in free water can be assessed with dip-slides. Small 'paddles', coated with nutritive agar gels, are dipped into any water sample, withdrawn and incubated. Colonies develop on the gels and are counted or estimated. This test can be used for sump bottom water or cooling water; it can be used for 'wet' oil but it will substantially underestimate the numbers of microbes present. There are several suppliers of dip-slides.

Tests for sulphate reducing bacteria on sump bottom water
In stagnant systems a semi-quantitative assay for SRB is essential; three suitable on-site tests for SRB are the Easicult 'S' test (Orion Diagnostica), the Sig Sulphide test (ECHA Microbiology) and the Sani-Check SRB kit (Biosan Laboratories

Inc.). In the Sig Sulphide test, a sample of bottom water is added to a glass tube of gel which is incubated for up to five days. If SRB are present, a black colouration progressively develops in the gel. Other test kits are broadly similar. These tests can also be used on 'wet' oil.

Sig Nitrite test on cooling water

This test semi-quantititatively detects nitrite reducing bacteria. These microbes are responsible for rapidly destroying nitrite corrosion inhibitors in engine cooling water. A sample is added to a glass tube of gel which is incubated for up to five days. A positive result is pink colouration and gas cavity formation in the gel.

This is a 'stand alone' test and can be used for problems in the cooling water systems. However, a positive result indicates that there is a potential source of contamination for the crankcase oil.

2.2.3 Interpretation

It is important that the temperature of incubation is comparable to the temperature of the system.

It can then be assumed that any microbes detected will be able to flourish at the system temperature. If the system temperature cycles or fluctuates it is preferable to run replicate tests at a range of temperatures. It will then be apparent which temperature is favouring the microbial growth. However if SRB are present they are likely to be active in static pockets of bottom water and a single incubation temperature (30-37°C) will be adequate. Whilst numerical values can be attached to warning and action limits for aerobic microbes, any detection of SRB is cause for concern.

Some suggestions for warning and action limits are given in Tables 1 and 2. These are a matter of opinion and should not be interpreted rigidly. An obvious trend to increasing numbers may be more important than actual numbers.

3. Anti-Microbial Measures

3.1 Crankcase Oils

3.1.1 Good Housekeeping

Good housekeeping emphasis is placed on water elimination and the correct operation of a purifier if present. Lubricating oils tend to entrain water, and this may only partially be resolved by the normal de-watering procedures of heating and centrifuging. Batch renovation, which involves heating the whole oil charge to about 80°C for one or two days, is less common than hitherto but is undoubtedly beneficial for its heat sterilising effect as well as its water separating function. However heat losses from the walls and bottom of the tank usually prevent complete heat sterilisation. Considerable data on heating as a key procedure in 'good-housekeeping' have been published[6].

Most large engines have a purifier to remove water and this continuously heats and centrifuges a slip-stream of the main lubricant charge and returns it to the crankcase. The purifier has been designed to reduce the viscosity of the oil by heating it and then to centrifuge it to remove water and particles. The purifier can have two important anti-microbial functions. At the centrifuge, many microbes which are entrained in water are removed with the water; others, suspended in the oil, are removed as far as the viscosity and retention time permit (the microbes have a density of about $1.05g/cm^3$). The most important anti-microbial effect can be at the heater. Both the contact time and temperature are important. Below 70°C there will be little pasteurising effect but between 70°C and 80°C there can be a significant kill; the contact time needed at 70°C is 30 sec or more but only a few seconds are needed at 80°C. Typical results are given in Figure 5. There will always be some survivors and these have the potential for further growth when they return to the main system. The purifier should be operated so that microbes can be killed there at least as fast as they can grow in the main system, and hence there is an important relationship between oil flow rate through the purifier and total oil volume. The mathematical basis for the computation has been given by Hill & Genner[6]. In general the flow volume per hour should be more than 10% of the total volume and preferably 20%. It is

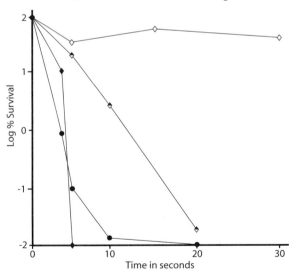

Figure 5. Survival of bacteria in infected crankcase oils when heated at various temperatures for up to 30 seconds. (After Hill and Genner[5]).
◇ Control, ◆ 60°C, ● 70°C, ◆ 80°C

important that the purifier suction is as low as practicable in the sump bottom as this is where growth usually starts.

Good housekeeping should extend to the spare oil tank and condensate water should be regularly drained off.

However, good housekeeping may lapse when a ship is laid up, and it is advisable to check for microbial contamination before lay-up and before re-commissioning. Dehumidification suppresses potential microbial growth.

High-speed engine lubricants can not suffer microbial spoilage in continuous use as the oil temperature achieved effectively pasteurises the system. A system may be at risk during lay-up.

3.1.2 Preservation

Good housekeeping utilises simple physical procedures to minimise microbial growth but if these fail, or are considered to be inadequate, the use of anti-micro-

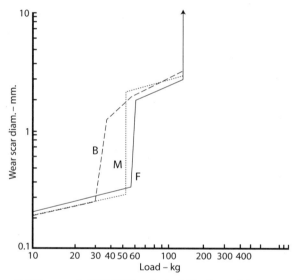

Figure 6. Four-Ball Wear results (IP 239/73) on fresh oil (F), oil containing an oxazolidine biocide (M) and oil containing a phenolic biocide (B). The oil with phenolic biocide only carries half the load which oils F and M can carry before a major increase in the wear rate occurs. (After Hill[1]).

bial chemicals (biocides) must be considered.

These are normally used at a high concentration to decontaminate fouled systems or at a lower 'preservative' concentration to protect clean systems from occasional minor exposure to microbes.

Any biocide used must be selected with great care as it is imperative that it does not impair the functional properties of the lubricant. In brief it must not affect the 'wear' properties of the lubricant or its water-shedding capabilities and it must be compatible with lubricant additives. Some procedures for carrying out appropriate tests have been given by Hill[4]. Figure 6 shows typical Shell Four-Ball Wear test results (IP 239/73). Experience has shown that the use of anti-microbial chemicals as preservatives is of limited success in engine oils. There is often an inherent chemical incompatibility and biocide deactivation which restricts the anti-microbial life of a stored preserved oil. When a preserved oil is put into use, the biocide may be thermally unstable or be washed out of the oil by alternating water ingress and removal. Preservation of hydraulic and stern bearing oils may be more successful and will be discussed later.

3.1.3   Decontaminating Oils in Use

Engine oil infections are often detected at an early stage by onboard testing, possibly after minor indications of malfunction. Experience has shown that growth of microbes in the lubricating oil is at first confined to a water pocket at the bottom of the sump. It can usually be successfully treated by pouring a biocide, possessing water and oil solubility, down the sounding pipe of the sump. The amount of biocide is estimated from a knowledge of the likely numbers of microbes in the pocket and the volume of the pocket.

Any biocide application to oil which is already heavily contaminated will result in the release into the oil of dead microbial sludges, which may block filters. Frequent filter changes must be anticipated until these dead sludges are removed. Supplementary purification may be practical to remove the sludges at the purifier centrifuge.

Where the ship is experiencing some severe malfunction and microbes have been detected throughout the oil system, it may be considered necessary to treat the whole of the oil charge in use with an oil-soluble biocide. If the ship is at sea the problems caused by dead microbes must be considered.

Continuous recirculation through the purifier to decontaminate an oil charge is always beneficial but unlikely to be completely effective. In an emergency, heavily contaminated oil can be passed slowly through the purifier to a spare clean tank and then returned to the main sump, if possible after this has been manually cleaned.

Occasionally the spare oil tank becomes infected; purification or biocide treatment (or both) are options.

It is obvious that a range of anti-microbial measures could be deployed, both physical and chemical, and expert help might be necessary to devise and implement the most appropriate and safe strategy. The suggestions which have made for complex problems with crankcase oil can be extrapolated to most other oil systems.

3.1.4   Decontaminating Systems Not in Service

At an early stage it must be decided whether the lubricating oil charge is to be dumped or retained as still serviceable. This decision may be based on analysis by the oil supplier or a contract laboratory. If it is to be retained, a procedure would be to pump most of the charge up for batch heat sterilisation, add oil-soluble biocide to the remainder of the charge and circulate this in the engine for 12 to 24 hours. This would then be dumped, and hand cleaning carried out. The heat-sterilised oil would then be returned to the engine via the purifier and topped up, and re-spiked with additives if considered necessary. Heat sterilisation in a renovating tank must be prolonged and rigorous (more than 80°C) as heat losses from the tank walls and bottom reduce the efficacy of this procedure. Bottom oil may not in fact be sterile and some should be drained off to waste with the water and sludge. If the oil has deteriorated substantially and is considered unserviceable, the whole oil charge can be heavily dosed with an oil-soluble biocide and circulated to sterilise the system before disposal and hand cleaning. It is much preferred to carry out these procedures with hot oil. Depending on the cleanliness achieved, the circulation of a flushing oil may be desirable. In high risk situations, biocide should be added to the new oil charge but the protection is likely to be temporary.

In the majority of incidents, microbial proliferation would also have taken place in the engine cooling waters (cylinder and piston), and leaks into the oil are major sources of both contaminating water and degradative microbes. These systems should therefore be emptied when convenient, flushed with aqueous biocide, hand cleaned and refilled. The cooling waters are invariably treated in use with corrosion-inhibiting chemicals, particularly nitrite-borate formulations or soluble oils, and these support microbial growth. Biocides cannot be used in coolants in operational use if the coolant is the heat source for the potable water evaporator.

3.2   Hydraulic Oil Systems

The strategies outlined in 3.1 may not be appropriate to hydraulic oil systems. Biocide additions made to systems in use via the header tank will have only a local effect if the system oil does not circulate through the tank; in most cases it would be unwise to add biocide to a contaminated hydraulic oil until the oil can be pumped out. After pumping out, a biocide flush of the system may be necessary. In complex systems microbial growth may be focused at 'slugs' of contaminating water. Even if there is a preservative biocide in the oil, its diffusion across the oil/water interface into the water 'slug' will be very slow under static oil conditions. Nevertheless a very careful biocide selection procedure, based on laboratory experiments, will indicate the most appropriate chemical preservative agent. Preservation is preferable to decontamination and Hill and Hill[7] proposed the following selection criteria for a hydraulic oil preservative:

- Soluble in and compatible with the hydraulic oil.
- Soluble in (sea) water.
- Long term biocide stability in oil and (sea) water.
- Broad anti-microbial activity at slightly alkaline pH.
- Acceptable health and safety/environmental impact.
- Preferably, amenable to on-site monitoring.

3.3   Other Straight Mineral Oils

Microbial growth in turbine oils is not uncommon, particularly in electricity generating units on land and on offshore platforms, and the treatment procedures would generally follow those suggested for ships' lubricating oil, always paying due regard to minimum interference with the functional properties of the oil. The water ingress is usually from steam condensation.

Rust preventive oils often pick up infected water from residual wash water on the metal components during manufacture. Strict control of the quality of the wash water is often the best way of preventing local growth and associated corrosion in the rust preventive oil film. For additional security, an anti-microbial chemical can be added to the rust preventive oil during manufacture. Onboard oiled spare parts which are rusting may have been affected by microbes growing in the oil coating; they may have to be degreased and re-oiled.

Stern shaft lubricants pick up sea water, which stimulates microbial growth. The oil is often supplied in a formulation which contains an anti-microbial chemical.

4.   Biocides for Oil

There is a very restricted choice of active agents available as few chemicals have the desirable characteristics referred to in section 3.1.2. Compatibility with the oil product is paramount; the cost and time required to determine this are such that there is little incentive to move away from tried and tested products. Those active chemical agents most used have been:
- NN methylene-bis (5-methyl oxazolidine).
- 4-(2-nitrobutyl)-morpholine.
- Isothiazolin derivatives.
- Thio cyanomethyl thio-benzathiazole.

They are usually formulated with other chemicals which modify their stability, solubility etc and these formulations are sold under a variety of trade names for a variety of applications. Some of the formulating chemicals, for example surfactants and solvents, may be undesirable in oils. All biocides are toxic and the product data sheets will advise on handling and disposal.

5.   Regulatory Matters

The microbial infection of lubricants is unlikely to pose a health hazard if protective clothing, including gloves and masks are worn and the creation of inhalable aerosols is avoided. The greatest risk will occur when manually cleaning out microbial sludges. Two particular microbes which are occasionally found, *Pseudomonas aeruginosa* and *Aspergillus fumigatus*, are opportunistic pathogens (they can infect people who have a low resistance to them); the former can infect cuts and the eyes, and the latter the lungs. They can only be identified in a competent laboratory which should provide professional advice if these organisms are present. A risk assessment may be advisable although it may not be mandatory.

If biocides are used for any purpose, a risk assessment will be advisable for their use, and any waste regulations must be complied with for discharge at sea or to shore. The supplier's precautionary measures must be followed. A health and safety risk will arise when the biocide concentrate is being handled. The concentration after mixing into the oil should not exceed the supplier's maximum recommended dose as this will have been set to avoid a health risk from this oil. Because oil biocides partition preferentially into any associated water, this water may eventually accumulate a hazardous biocide concentration. Therefore, water pumped out from a sump bottom or drained from a renovating tank or purifier centrifuge must be handled and discharged with appropriate precautions.

Biocides supplied in America will be registered with the Environmental Protection Agency which will have demanded a dossier satisfying health and safety and environmental concerns. In Europe the EU Biocidal Products Directive 98/8/EC will be progressively promulgated by national authorities; it will have a similar impact as the EPA regulations but it may not be fully effective until 2008. Some good European biocides are not acceptable in North America and vice versa.

The disposal of biocides requires that they are diluted until they are not toxic to marine/river life or that they are chemically deactivated before they are discharged (various MARPOL regulations). If a contractor takes away waste containing a biocide he must be informed of its presence; the onus for safe disposal remains with the person who generated the waste. Biocides used on offshore platforms are listed and categorised in the DTI's Revised Offshore Chemical Notification Scheme 1997; this stipulates maximum permitted discharges. Engine oil containing biocide can be disposed of by combustion.

6.   Future Trends

Without doubt, over the last two decades the incidence of microbial problems in lubricant and hydraulic oils has decreased. In some part this has been due to better monitoring, better purification and early recognition, but a significant factor has been the development of oil products more resistant to microbial spoilage. There is unfortunately a backlash to this latter trend as a vociferous lobby, worried by visible persistent signs of oil spillages, is demanding that oil products should be biodegradable. There is thus a paradox, how can biodegradable oil products retain a resistance to biodeterioration in use. This is as yet unanswered, but one approach is to preserve them with biocides which can be readily neutralised when the product reaches the end of its useful working life. The small size of the market for oil biocides is not conducive to prolonged and expensive research in this area. Fortunately in-line heat treatment for the crankcase oil is normally effective as an avoidance measure. For other oil systems, intermittent heat treatment may have to be considered; this is fairly successful for straight oils used to lubricate metal working operations.

7. References

1.   Hill, EC, (1975). *Biodeterioration of Petroleum Products.* in Microbial Aspects of the Deterioration of Materials, eds. EW Lovelock and R. J. Gilbert, Academic Press, 127-136.

2.   Hill, EC and Al-Haidary NK, (1976). *Some Aspects of Microbial Corrosion in Rolling Mills.* Proc. 'Microbial Corrosion Affecting the Petroleum Industry, Inst. Petrol., London, IP 77-001.

3.   Hill, EC, (1978a). *Microbial Aspects of Corrosion, Equipment Malfunction and Systems Failure in the Marine Industry.* Technical Research Report TR/069. General Council of British Shipping. Reprinted 1983 as TR/104.

4.   Hill, EC, (1978b). *Microbial Degradation of Marine Lubricants – Its Detection and Control.* Transactions of the Institute of Marine Engineers. 90, 197-216.

5.   Hill, EC and Genner, C, (1980). *Ecological and Functional Aspects of the Microbial Spoilage of Marine Engine Lubricants and Coolants.* In Biodeterioration, eds. TA Oxley, G Becker and D Allsop. Pitman Publ. Ltd., London, 37-43.

6.   Hill, EC and Genner, C, (1981). *Avoidance of Microbial Infection and Corrosion in Slow Speed Diesel Engines by Improved Design of the Crankcase Oil System.* Tribology International. 14, 67-74.

7.   Hill, EC.and Hill, GC, (1996). *Prevention of Microbiological Growth in a Sub-sea Hydraulic System.* 10th International Colloquium, Tribology, Technische Akademie Esslingen, 2223-2228.

8.   Hill, EC, (1996). *Microbes and Shipping.* Marine Consultant and Surveyor, Vol.4, 39-42.

9.   Hill, GC and Hill, EC, (1997). *Microbiological Quality of Fuel – Trends Tests and Treatments.* 1st International Colloquium, "Fuels". Technische Akademie Esslingen 16-17 Jan 1997, 269-273.

10. Hill, EC, Collins, DJ, and Hill, GC, (1999). *Microbiological Monitoring On-Site.* Proc. Int Conf. On Condition Monitoring, Swansea 12-18 April 1999, 11-25.

4. Microbial Problems in Bilge and Ballast

GC Hill

Contents

1. The Marine Industry and Microbially Influenced Corrosion (MIC)

Microbially influenced corrosion (MIC) is a well documented phenomenon which has affected the marine industry spasmodically. Very severe pitting corrosion has occurred in bilges, most notably in vessels operating inshore, and in ballast tanks. More recently, rapid and severe pitting corrosion in cargo tanks of crude oil and petroleum products tankers has been attributed to microbial activity. Microbial proliferation can influence and accelerate normal electrochemical corrosion processes, particularly when microbial consortia which include sulphate reducing bacteria (SRB) are involved. The mechanisms involved have already been discussed in Chapter 1 section 5 and will not be described in detail again here. Relative sections in the introductory chapter will be referenced where appropriate. Iron and steel, including stainless steel, are affected as are aluminium and copper and their alloys. Nickel alloys may also be susceptible, although there is little evidence which conclusively points to a microbial involvement where corrosion has been observed. Titanium is generally believed to be resistant to MIC. Estimates have been made that at least 10% of all corrosion incidents have a microbial involvement enabling tentative assessments of the costs of MIC. In the USA alone, these exceed \$16 000 million a year in industry as a whole. The marine industry is not alone in its susceptibility to MIC but it is particularly susceptible due to the frequent proximity of steel structures to sea water which provides sulphate for SRB and it is often contaminated with organic nutrients and corrosion influencing microbes. This chapter will in particular review MIC in bilges and ballast tanks. The factors affecting that corrosion will be discussed as will the techniques which can be used to help identify, monitor and prevent MIC. MIC in cargo and fuel tanks will also be discussed.

2. The Microbiological Process

As in every microbial process there are certain essential pre-requisites for growth, namely water, food and a suitable physical environment. The processes in bilges and in ballast tanks are discussed separately below, although there are similarities, notably the role of sea water which contains essential nutrients that, if absent, would effectively limit microbial activity. The sulphate reducing bacteria (SRB) are linked to most marine incidents of MIC although in some special circumstances, which will be dealt with separately, entirely different microbial types are involved. It must also be emphasised that although SRB are often considered the bête noir of corrosion problems, they almost invariably act within consortia of many microbial types. Those parameters which optimise microbial proliferation and corrosion must relate to the entire consortium rather than any individual species.

2.1   MIC in Bilges

It has been speculated for many years that bilge water will provide an ideal environment for microbial proliferation and that consequently internal MIC of hull plate could be a concern.

In a study by Hill in the late 1960s, a survey of 37 vessels, mostly tankers and ferries, was conducted by testing a total of 78 water samples, from various locations within the bilges, for the numbers of aerobic bacteria, yeasts and moulds and for the presence of SRB. The pH was also tested.

The survey did not extend to correlating microbial contamination with corrosion history but confirmed that extensive microbial proliferation could occur although only about 20% of the samples were found to contain SRB. There was substantial variation in the types of organisms present from location to location within the bilges. It was believed that in most cases the bilge water in these vessels was a result of ingress of offshore sea water. Studies were only resumed when, in the early 1990s, a spate of severe hull pitting corrosion incidents were reported, predominantly in the bilges of vessels operating inshore. Investigations were undertaken to determine whether microbes were involved and what were the factors which had resulted in the onset of so many incidents. Subsequent surveys, particularly of tugs, ferries and dredgers, revealed that a significant proportion of bilge water samples from inshore vessels contained large numbers of SRB. The presence of heavy SRB infection correlated with the incidence of severe corrosion and hence they were implicated as the principal cause. Again, substantial variation in microbial types with sampling location was noted. Although much of our understanding remains undocumented and is based on anecdotal evidence and intuitive speculation, a number of important factors have been identified.

It is not hard to envisage how microbes will enter the bilge. Bilges tend to be the dirtiest part of the vessel and dirt is synonymous with microbes. Microbes are ubiquitous in the environment and, where conditions are suitable, they can proliferate from a low background contamination level to many millions per ml. For a corrosive microbial consortium however, the right types of microbes must be introduced - those which can utilise any nutrients present and, of course, the SRB. The importance of nutrients is discussed below and the types of microbes which can readily degrade the nutrients typically present in bilges will not be uncommon in sea water. This is particularly true of inshore sea water which suffers chronic oil pollution and will contain a ready adapted population of hydrocarbon degraders; it will also probably contain nutritious fertilisers in run-off water from agricultural land. SRB are usually present in sea water in very low numbers, perhaps <1 per litre, but, it only requires one viable SRB to propagate a population of many millions if nutrient requirements are met and environmental conditions are conducive. Any microbially contaminated substances from the ship which enters the bilge, such as contaminated fuel and oil and sewage tank overflows, will be a further source of microbes. Within the bilge the different types of microbes will become established in their individual niches. While some will remain freely suspended in the bilge water, many more, including SRB, will attach to the internal surfaces of the hull plate as biofilm. It is the organisms in the biofilm, in close proximity to the steel, that are directly involved in MIC. Many microbial types may make up the biofilm and the interdependence of these is discussed further below.

There will be no shortage of nutrients for microbes in bilge water, particularly where cleaning is sporadic. Leakage of oil and other organic materials such as engine cleaners will provide an ample carbon source and this will be supplemented by overflows from sewage or grey water tanks; any detergents will help emulsify oil, making it easier for microbes to degrade. Sea water will provide a source of those essential nutrients which are required in smaller amounts and, importantly, will also contain sulphate for the SRB.

However, offshore sea water will contain relatively low concentrations of some nutrients, notably nitrogen and phosphorus which will probably be present at < 1ppm. When in short supply, these two elements limit the extent of microbial growth. When levels of nitrogen and phosphorus are increased by comparatively small amounts, dramatic increases in microbial proliferation are possible; progressive stimulation occurs as nitrogen concentration increases up to 11ppm and as phosphorus increases up to about 2ppm. Inshore sea water is likely to be richer in these essential nutrients and will also contain higher levels of organic material. The nutritive status will be further enhanced where sea water is chronically polluted with agricultural fertilisers; severe bilge corrosion has been observed in tugboats which moor adjacent to a dock routinely handling fertiliser causing chronic pollution of the dock water. Leaks of engine cooling water into the bilge may provide an additional source of nitrogen if a nitrite corrosion inhibitor is used.

The nutritive mixture in the bilges of inshore vessels will provide a balanced diet for a wide range of microbes but will not as yet be sufficient to support the SRB. SRB cannot directly utilise hydrocarbons such as oil; they require simpler molecules as a carbon source, typically organic acids, for instance succinate, lactate and acetate. For these requirements the SRB are dependent on the activity of aerobic hydrocarbon degraders which precede them. These hydrocarbon degraders produce soluble, partially-oxidised compounds which migrate throughout the water and in turn become nutrients for other microbes, including the SRB. Thus the proliferation of SRB is usually dependent upon the aerobic hydrocarbon degraders.

To the layman it seems a paradox that the SRB which are anaerobic organisms, intolerant of oxygen, will thrive alongside aerobic organisms. The explanation lies in the fact that in a typical bilge, oxygen concentration will vary both spatially and with time. In a stagnant bilge there will be a sharp decrease in oxygen concentration within a few centimetres of the surface because aerobic organisms in the bilge water will feed and respire until all oxygen has been utilised; many of these aerobes will then lie dormant until further oxygen is introduced from the air, for example when water in the bilge is agitated when the vessel puts to sea. Some of the organisms which degrade hydrocarbons aerobically can, in fact, continue to proliferate in the absence of oxygen by switching to an anaerobic metabolism; these are known as facultative organisms. It is while the bilge is stagnant that the SRB can thrive on the organic acids produced by the aerobic organisms. If a cycle of aeration and stagnation is perpetuated, corresponding to the vessel being in use and berthed, both the aerobic hydrocarbon degraders and anaerobic SRB can thrive. It is apparent that any prolonging or exacerbation of stagnation will promote the activity of SRB. Hence, an important factor in the recent increase in hull corrosion incidents has been identified.

Legislation which restricts the direct pumping of bilge to sea near the shore, means that those vessels which rarely traverse this boundary will tend to accumulate bilge on board. An alternative is to pump bilge to a shore waste contractor and, as this is expensive, there has been a tendency to delay pumping ashore until it is absolutely necessary. Consequently, large volumes of bilge water may accumulate and, because of the smaller relative surface area exposed to atmospheric oxygen, the bilge will be more stagnant. Prior to this legislation when bilges were pumped out regularly into the sea, stagnation was avoided and the oxygenated air/water interface was lowered to a level which would be inhibitory to SRB on the bottom plate. Additionally, regular pumping out of bilges will help to remove both nutrients and proliferating microbes.

Deep narrow bilges are likely to become particularly stagnant and vessels which accumulate water in this type of bilge have proven particularly susceptible to MIC. However, severe pitting corrosion observed on the bottom plate of a new dredger, testifies that a wide flat bilge will not be immune to MIC.

Vessels which operate only infrequently, for example stand-by tugs, will be prone to long stagnation of bilge water and it has been shown that these are amongst the most susceptible to MIC of hull plate.

Although the concept of a cycle of aeration and stagnation in the bilge helps illustrate a major factor in the corrosion process, this may be an over simplification. The real scenario is likely to be more complex as it has been shown that aerobic microbes and anaerobic organisms can thrive simultaneously in close proximity. This is particularly true in biofilms which are a heterogeneous mix of many microbial types embedded within a polysaccharide slime matrix stuck to the steel surface. A small micro-colony of aerobic hydrocarbon degraders within the biofilm may thrive directly adjacent to a micro-colony of SRB. The former will produce organic acids and cause localised depletion of oxygen allowing the latter to flourish. In fact, the establishing of such oxygen gradients may be an additional corrosion mechanism (see Chapter 1 section 5.2) to complement and enhance the more direct corrosion mechanisms of the SRB (see Chapter 1 section 5.5). However, where a biofilm is exceptionally thick, or where a thick sludge is present, there may be no oxygen penetration below a certain depth and, once organic nutrients produced by aerobic organisms are exhausted, SRB can no longer thrive. Thus, in locations where exceptionally prolonged anaerobic conditions occur, there may be no significant SRB activity. The influence of sludge thickness appears to be complex and difficult to predict but compact deep sludges have been seen to have a protective effect as opposed to being sites of heavy SRB proliferation. A certain degree of intermittent aeration would thus appear to optimise SRB activity.

A number of other ecological factors should be mentioned. It is well known that microbes like warmth and, with the exception of a few organisms with specialist temperature requirements, those organisms that grow well at say 10°C are likely to grow even better at 25 or 30°C. In bilges, the engine will provide warmth, and corrosion is frequently seen to be most prevalent and severe in bottom plate immediately below the engine, frequently below the flywheel or gearbox. The influence of temperature may thus be the reason.

The utilisation of oxygen in bilge water by aerobic microbes plus the electrode potential of steel hull plate will tend to cause a voltage gradient (redox potential (Eh) gradient). Eh will usually be positive at the air/water interface but negative in close proximity to uncoated steel plate. Sulphide generation by SRB and hence their corrosive activity, will be optimised when Eh falls to about -200mV because SRB only reduce sulphate to sulphide when redox potential is significantly negative; at less negative potentials SRB will tend towards the reduction of phosphates or nitrates, processes which do not result in the formation of significantly corrosive products. Microbial activity both influences and is influenced by Eh. As aerobic microbes utilise oxygen, Eh will drop. Negative Eh will promote the activity of anaerobic microbes and also facultative organisms (capable of aerobic or anaerobic growth).

There is some evidence that SRB influenced corrosion only occurs within a pH range of about 5 to 9 with an optimum being around 7.5. The hydrocarbon degrading bacteria may influence pH in bilges, for example by producing organic acids (pH lowered) or by producing ammonia (pH raised). However, because sea water is strongly buffered (resists pH change), in most bilges the influence of microbes on pH will be minimal and pH will remain just above neutral and at an optimum for SRB.

Localised pH variations may occur however; for example, under a microbial slime the conditions may be very acidic whilst overlying water is alkaline. pH changes can occur throughout the bilge in a vessel as a consequence of acid or alkali chemical contaminants in the water; in the author's experience, pH values as diverse as 5 to 9 or over may be encountered.

Consideration of the factors described above will help determine which vessels may be particularly at risk from MIC, so enabling an effective strategy of monitoring and prevention to be instigated. It is apparent however, that the factors which influence MIC are complex and still not completely understood. There are as many different manifestations of MIC as there are types of vessel. Typically SRB pitting corrosion is seen on bottom plate beneath engines but its occurrence and distribution is usually sporadic and it may occur away from the most obvious sites, perhaps, for example, influenced by unpredictable turbulence

in the bilge water caused by suction pumps. In some cases, severe pitting and perforation has been observed in the bilge suction pipes, where the design and operating conditions are such that the pipes do not drain and remain full of stagnant
bilge water. There is some suggestion that SRB pitting corrosion is more prevalent around welds, but again this is a phenomenon which is not completely
understood. The presence, type and condition of coating in the bilge is also a
very important consideration and this is discussed further in section 5.1.

2.2   MIC in Ballast Tanks

Microbially influenced corrosion in ballast tanks is also a spasmodic problem in
the marine industry. The factors discussed above relating to bilge corrosion are
by and large also applicable to ballast tank corrosion. Ballasting with dirty
inshore water, which will contain nutrients and degradative organisms, is likely
to give a higher risk of corrosion than ballasting with clean, open-sea water.
Leaks of oil or waste into ballast tanks will further stimulate microbial corrosion. Permanently-full ballast tanks are possibly less affected than ballast tanks
where there is a high turnover of water and hence also of nutrients. In permanent ballast tanks, microbial processes will only go as far as the limited availability of nutrients (eg sulphate) and oxygen allows. Temperature may be an
important factor in ballast tank corrosion, as illustrated by one case-history
involving severe pitting corrosion in ballast tank pipes which passed through a
heated fuel oil tank. Once the pipes had become perforated, leaks of fuel oil into
the ballast tank probably further aggravated the microbial corrosion process.

It is possible that in ballast tanks, types of microbially influenced corrosion
other than that involving SRB may be important. The role of sulphur oxidising
bacteria (SOB) in corrosion in ballast tanks is not fully understood but it is feasible that they recycle sulphide produced by the SRB and, in the process, generate corrosive sulphuric acid. SOB are aerobic and will only oxidise sulphur compounds if oxygen is available.

However, the species implicated in corrosion (aciduric species) only proliferate when pH is already acidic (eg below pH4) and in the presence of oxygen.
Consequently, in ballast tanks they will not be active in bulk sea water which is
highly buffered against pH change and which will neutralise and dilute any acid
production. Nevertheless, the SOB can proliferate in damp - as opposed to wet -
conditions, and thus may play a role in deckhead corrosion, living in condensate
on the tank walls and deckhead. Here they could oxidise sulphide and other sulphur compounds that dissolve in the condensate. In condensate there will be no
buffering capacity, and localised highly-acid conditions may develop, resulting in
corrosion. SOB have been found on wall and deckhead plate in the voids above
crude oil, and very local strong acid has been detected. Inert gas usually contains
sufficient residual oxygen to sustain aerobic microbes and their presence is therefore not surprising. However their role, if any, in accelerating corrosion has not
yet been substantiated.

As in most MIC incidents, it is probably the case in ballast tank corrosion that microbes exacerbate or work alongside other non-microbial corrosion processes.

2.3 Health and Safety and Environmental Issues of Microbes in Bilge and Ballast.

Wherever SRB are active in an enclosed space, of considerable concern is the potential danger from microbially produced hydrogen sulphide. Hydrogen sulphide is more toxic than hydrogen cyanide and it causes serious distress and fatalities every year. Up to 300ppm concentration in air is at first unpleasant and then causes increasing dizziness. At concentrations in air above 700ppm, hydrogen sulphide is rapidly lethal. Unfortunately hydrogen sulphide anaesthetises the sensory organs and a false sense of security is given if the offensive smell disappears. In a closed or virtually closed tank space, hydrogen sulphide will equilibrate between the water- and air phases such that even a few ppm in the water phase will equilibrate with a lethal concentration in the air phase. Up to 50ppm water phase concentrations have been recorded due to SRB proliferation in oil platform legs; a few ppm in infected tank water is not uncommon. SRB can be regarded as small, living gas-generators that will go on generating hydrogen sulphide until they are killed. A tank can be gas tested and found free of hydrogen sulphide, but subsequent disturbance of sludge (eg during tank cleaning) can stimulate previously dormant SRB, causing dramatic and unpredictable increases in hydrogen sulphide concentration. In bilges at the bottom of a well ventilated engine room, SRB are unlikely to pose any serious health and safety concerns but may be responsible for poor working conditions by creation of exceedingly unpleasant sulphurous odours, usually first noticeable when the vessel puts to sea. It is apparent that the hazard posed from microbially generated sulphide should form part of any health and safety risk assessment prior to entering ballast tanks.

3. Recognition and Monitoring

3.1 General Indicators of Microbially Influenced Corrosion

Where operating conditions are such that some of the factors discussed above are optimised for SRB corrosion it will be prudent to instigate some form of monitoring strategy.

There may be clues that MIC problems exist, for example sulphurous odours emanating from the bilge, particularly when the water is disturbed when putting to sea. Occurrence of severe pitting corrosion or perforation of hull bottom plate would be strongly indicative of a microbial problem. The size and shape of pits can give a clear indication that microbes have been involved in their formation.

SRB pits typically, although not always, have terraced sides as indicated in

Figure 1. When soft corrosion deposits are first removed, the inside of pits tend to have a dull grey or even slightly shiny appearance. Black sulphide around the corrosion site is another indicator of SRB activity.

Figure 1. Steel 10mm bottom plate (approx 13cm x 20cm)
exhibiting typical pitting corrosion caused by SRB

No analysis, microbiological or otherwise, can give a conclusive indication that an observed corrosion incident has been influenced by microbes. The multiplicity and complexity of possible corrosion processes excludes the possibility that a single simple test can give conclusive proof that MIC has occurred or is likely to occur.

Nevertheless a few simple tests can be conducted on-board; the results of these enable the operator to make decisions about whether it is appropriate to instigate anti-microbial measures. Three parameters which are easily monitored are:
- microbial sulphide generating activity.
- redox (electrode) potential (Eh).
- pH.

3.2   Monitoring Microbial Sulphide Generation

3.2.1   Selection of Tests

Perhaps the most important characteristic of the sulphate reducing bacteria in creating a corrosive environment is their ability, usually with the co-operation of other microbial species, to generate sulphide. A number of cheap and exceedingly simple tests are available which enable the monitoring of microbial sulphide generating activity onboard. Typically these consist of a glass tube containing a nutritive gel which is 'incubated' after adding a sample. The sample is added either by stabbing it into the gel using a glass capillary tube (Easicult S Test, Orion Diagnostica) or a swab (Sanicheck SRB, Biosan Laboratories Inc.) or more simply by pouring a sample directly onto the gel (Sig Sulphide Test, ECHA Microbiology Ltd.). If sulphide generating bacteria are present in the sample, with incubation, the gel will turn black; usually this will happen overnight for a heavy infection, or after two to seven days for moderate or light infections. The extent and rate of this blackening are approximate indicators of the severity of infection. These types of tests offer a considerable advantage over standard laboratory tests for SRB such as NACE TMO-194-94 which, although perhaps more specific for SRB, can take anything up to 28 days incubation before a full result is available. Although tests such as the Sig Sulphide, Easicult S and Sanicheck S tests need incubation, there need be nothing sophisticated about the way in which this is done. Any warm location of between 25 - 30°C, for example in the engine room, will suffice. Incubation at temperatures in excess of 35°C should be avoided unless specifically investigating infection in a warm system of comparable temperature, for example in the bottom water of a crude oil tanker.

3.2.2 How and When to Sample

Sample bottles should ideally be sterile. Suitable disposable bottles can be purchased from laboratory or hospital suppliers or from the suppliers of the test kits. Some sample bottles (eg the Dippa) have a detachable handle which aids sampling from awkward locations.

The ideal sample for detection of sulphide generating microbes contributing to corrosion, will be one taken from close to the potential site of that corrosion. In a bilge, this will typically be from the deepest part underneath, or just behind, the engine. Unfortunately this is not the easiest of sites to sample but, with a little innovation, a sampling device can usually be constructed, typically a sample bottle taped to the end of a broom handle. Purpose-made bottom samplers such as those employed for sampling liquid cargo tanks could be used if a more sophisticated approach is required. When conducting an initial investigation it is a good idea to sample from several locations around the bilges. The sporadic and sometimes unpredictable way in which SRB distribute has already been alluded to, and sampling several locations in an initial investigation enables any 'black spots' for infection to be identified. The material sampled is also important; a sample of waste oil floating on the surface of the bilge is not appropriate for analysis. A sample of water phase is more appropriate and if this can include some sludge from the bottom, all the better. Where a sample is not easily obtained directly from the bottom of the bilge, one can be taken using the bilge suction pipe. The bottle should always be filled completely.

In ballast tanks, sampling may be more problematic. A sample from the tank bottom is ideal. For some tanks a bottom sampling device can be used, but in others an appropriate sample will only be available when the tank is open and prepared for manual entry. In such a case a sample of sludge or water slops from the tank bottom can be taken but it should be ensured that, as far as possible, this relates to the contents of the tank prior to emptying. Once the tank is emptied and ventilated, water will soon become aerated and therefore samples for analysis for anaerobic sulphide generating microbes should be taken at the earliest opportunity. Any slops water should be residual from the previous ballast and should not be remains of cleaning or flushing water. As a tank dries and ventilates, sulphide generating microbes will slowly die and so the importance of maximising opportunities for sampling and analysis of ballast tanks cannot be overstressed. Because there tends to be only minimal opportunities to take suitable samples from ballast tanks, it would seem prudent to sample and test whenever tanks are opened as part of an inspection routine. If tank bottom samples cannot be obtained, a sample of discharged water from the tank is better than no sample at all.

There are no hard and fast rules about the frequency of taking routine samples for monitoring for sulphide generating microbes. The prudent operator of a vessel known to have been subject to severe MIC in the bilge, might consider a monthly onboard test appropriate. A yearly sample might be sufficient for a situation which is considered low risk. It is a surety, however, that the complete absence of a simple monitoring regime is the best possible way to expose a vessel operator to a highly-expensive corrosion incident, as some have found to their cost. The essence of routine monitoring is to detect the potential for a corrosion incident and, if necessary, prevent it before it actually happens.

It is really down to the operator to decide, with expert advice if necessary, on a frequency of monitoring which will give confidence that development of any corrosion problems can be remedied in a timely fashion.

Even the most severe of corrosion problems do not develop overnight, but over a period of several months. The factors listed in section 3.4 below will also help determine the perceived risk of MIC and thus help dictate an appropriate monitoring frequency.

When investigating a corrosion incident subsequent to an inspection, it may be difficult to obtain suitable valid samples for microbial analysis if more than a few days have elapsed since removal of sludge and water from the affected area. Nevertheless, we have frequently been surprised at the survival time of SRB in an open, empty tank, particularly if they are shielded in scale or corrosion deposits. We have even detected SRB on corrosion scale samples analysed several months after their removal from corroded steel. The key factor in the 'die off' of SRB in such sample material seems to be desiccation rather than aeration; thus if a sample can be sealed to keep it moist, analysis some weeks later may still provide useful data. When investigating a corrosion incident, a Sig Sulphide, Sanicheck SRB or Easicult S test, conducted by swabbing the inside of a still damp corrosion pit and stabbing the swab into the gel, will probably give the best possible indicator that SRB may have been involved in the pit's formation.

As already stated, the sooner the test is conducted after exposing the corrosion, the more reliable the result will be. As time passes, quantitative data obtained becomes less relevant to the conditions which were present at the time corrosion was occurring, but tests may still provide qualitative data about the microbes present (ie presence/absence of SRB). Failure to detect sulphide generating microbes in a pit or corrosion scale which has had prolonged exposure to air would, of course, not exclude the possibility that they may have played a role in the corrosion process.

3.3   Monitoring Eh and pH

Redox potential (Eh) is easily monitored by cheap hand-held meters which are dipped into a water sample and give an instant readout. The simplest and cheapest of devices is appropriate for onboard applications. Conducting an Eh test is a useful adjunct to a test for sulphide generating microbes, and interpretation of results is discussed in more detail in section 3.4 below. It has already been stated that, in practice, microbes are unlikely to exert any great effect on the pH of sea water and that where they are responsible for acidic corrosion, pH changes are highly localised. Thus while pH measurement is unlikely to provide any useful indicator that MIC is occurring, an occasional test can help establish whether optimum conditions for MIC exist. Cheap hand-held meters can again be used but even-cheaper pH indicator paper will suffice. Paper can be pressed onto a defined area of corrosion as an indication of local pH. Eh meters and pH indicator papers are available from laboratory suppliers and some test kit suppliers.

3.4   Interpretation of Tests

It is likely that the majority of, if not all, MIC incidents in bilge and ballast involve the microbial generation of sulphide. Thus it is reasonable to assume that a regular negative result for microbial sulphide generation, obtained by a simple test, is a good assurance that the tank or system is unlikely to be subject to MIC.

Attention is drawn, however, to earlier comments regarding the relevance of the sample tested (see section 3.2.2 above). Detection of microbial sulphide generation is not necessarily proof that MIC is occurring.

However, an approximate determination of the 'risk' that a tank/system is prone, or may soon be prone, to MIC can be achieved by consideration of the result of tests for sulphide generating microbes in conjunction with the results of pH and Eh tests and the following factors:

- The source, type and cleanliness of water in the tank or bilge.
- Whether a coating is present and if so whether it is in good condition.
- The frequency of turnover of water in the tank or system.
- The volume of water in the tank or system.
- The overall cleanliness of the system, in particular the presence of organic substances such as waste oil, sewage, run-off of engine room cleaning agents,

glycols and also corrosion inhibitors derived from cooling water leaks.
- The amount of sludge accumulation.
- Temperature.

The implications of each of the above factors have already been discussed. It is usually the experience of this author that where exceedingly-high SRB numbers and/or 'heavy' microbial sulphide generation is detected, other factors will usually be optimised for SRB influenced corrosion.

When monitoring the Eh of a bottom sample from bilge or ballast, the engineer should consider a result approaching or exceeding -200mV to indicate a high risk of MIC if significant microbial sulphide generating activity has also been detected. A less negative result should still be treated with concern. A positive Eh in water from the bottom of a tank or bilge suggests SRB influenced corrosion is unlikely to be occurring at the time of sampling, but one should always consider that any substantial aeration of the bilge or sample prior to testing may cause temporary Eh increases.

Where pH deviates significantly from neutral (eg by one or more) the risk of MIC by SRB can be considered to be lower. Where pH is outside the range 5 to 9, SRB influenced corrosion is unlikely. Any corrosion caused by microbially generated acids will generally be due to highly localised acidity on the steel surface and not easily detectable by a pH test of bulk ballast or bilge water. The dramatic pH drops associated with corrosion by acidophilic sulphur oxidising bacteria are generally not encountered in bilge. In ballast they will only be detectable in damp, as opposed to wet, conditions, for example on steel surfaces above the normal fluid level.

Attempts have been made to correlate numbers of sulphate reducing bacteria, as indicated, for example, by a laboratory test in accordance with NACE TMO-194-94, with a corrosion risk. In truth correlations are difficult to establish. The most likely reason is that the numbers of SRB detected in a water sample, even if it is drawn from close to a steel surface, will not necessarily be influenced by the numbers of SRB actually attached to the steel surface at a potential corrosion site. In the author's opinion, however, some very broad assumptions can be made. Very heavy SRB contamination in a water sample (for example 10^6/ml) is probably indicative of a higher risk of corrosion than a light contamination (10^2/ml). Values in between might be indicative of moderate risk.

Onboard tests generally do not give a direct assessment of number of SRB but instead give an approximation of the extent of microbial sulphide generation (according to the amount and rate of blackening in the test gel); this is perhaps a more reliable assessment of corrosion risk than a numerical value of SRB. Guidance on interpretation will usually be given in the test instruction. The Sig Sulphide test instructions include a graph that allows the result obtained to be approximately correlated with that which might be expected had a NACE TMO-194-94 type test been conducted on the same sample.

For those more familiar with interpretation of number values for microbes, this can be useful, but slight variations in results should not be considered to be significant. A distinction in terms of absent, light, moderate or heavy infection, is probably the best that can be expected for onboard tests.

4.   Remediation.

Ultimately, where a problem of MIC exists, it will be usually necessary to employ some type of anti-microbial chemical treatment (biocide) to kill the microbes involved. Physical cleaning alone will not prevent rapid redevelopment of the microbial populations which stimulate the corrosion process. The principles of using chemicals to control microbes have been outlined in the introductory chapter of this book (Chapter 1 section 4) and it is not intended to provide a detailed review of all appropriate chemicals here. A few factors which must be considered in devising a safe and effective strategy are worth reiterating.

Once MIC has started, it cannot always be stopped simply by adding anti-microbial chemicals. These cannot penetrate biofilm, sludge or mud at normal, safe, in-use concentrations; microbes entrained therein will survive. Once ferrous sulphide has formed, it will continue to act as a cathode even though the SRB which generated it have been killed. Cleaning is a pre-requisite for successful microbiological decontamination and it also removes aggressive corrosion products.

Whatever is used for the decontamination must be safe for the operators and the environment. An additional consideration for bilges is that the biocide must not interfere with the effective operation of the oily-water separator. Use of liquid biocide concentrates in bilges is not ideal. Although in most cases, once biocides are diluted to recommended use, concentration hazards to health are insignificant; there may be considerable difficulties handling the concentrates safely in an engine room. Rarely will it be possible to accurately assess the volume of water in a bilge and, consequently, it may not be easy to determine the amount of biocide concentrate which should be added. Overdosing can present a health hazard. Underdosing can result in ineffective decontamination. Different biocides are active to different degrees, and there is no universally-effective concentration; factors such as bilge cleanliness and extent of contamination will play an important role in determining the effective concentration. Generally, advice from the biocide supplier or independent consultants is prudent. If, as is usually recommended, the bilge is emptied and cleaned prior to treatment, water will have to be deliberately added with the biocide in such a way that a good mix of biocide is obtained. Thorough mixing and dispersal of biocides is necessary for effective decontamination.

If areas of the bilge do not receive an adequate dose, the organisms which survive there will re-colonise the bilge once the biocide disappears. Particular attention should be paid, for example, to isolated compartments and to pipes which may retain contaminated water.

Some biocides have very strong odours; if applied, for example, to an open engine room bilge, the odour can make subsequent entry to the engine room at best unpleasant and at worse not feasible until the odour has dispersed.

No biocide works instantaneously and generally a contact time of around 24 hours is required, longer if the biocide has to penetrate residual sludge. Effective decontamination can rarely be achieved quickly and, if the treatment process prevents normal operation of the vessel, decontamination can be costly. It does not necessarily follow that biocides which are active at lower concentrations will also be faster acting; some biocides are very effective but so slow acting that they are not economically viable options for bilge decontamination.

An alternative to the traditional broad spectrum biocide is to use an SRB-specific biocide. Such a biocide has been developed with bilge treatment specifically in mind; it can be easily and safely applied as a powder contained in water-soluble sachets. Sachets are distributed around bilge compartments. Overdosing is not a concern as, even in the powder form, the biocide is relatively safe to handle. It is sparingly soluble, so once added to the bilge it sinks to the bottom and disperses slowly, aided by vessel motion, giving a relatively long-term protection. Most importantly, treatment need not interfere with normal vessel operations

For large systems or tanks with a high turnover such as ballast tanks, chemical costs will be an additional factor in the selection of the appropriate agent for decontamination. Cheap halogen-based oxidising biocides (similar to normal household bleach) are then usually more cost effective than specialist products. They can be easily monitored in use to ensure an adequate concentration is present and are easily detoxified before discharge. A disadvantage is that they are rapidly inactivated by even small amounts of dirt and organic material and so thorough cleaning is required before treatment. Halogen biocides are usually not practical for bilge treatment because thorough cleaning is not easily achieved.

5.   Avoidance

The old adage 'prevention is better than cure' is most relevant to the control of MIC in bilges. The potential costs of drydocking, decontamination and steel replacement far exceed those of a sensible routine monitoring and treatment programme. It is almost a certainty that where a vessel has been subject to microbially influenced corrosion, the problem will return unless an ongoing avoidance strategy is implemented. This may simply be a change in operating procedures, for example more frequent discharge of the bilge water. A more thorough strategy would involve an onboard monitoring strategy. Prompted by regular testing, or on the assumption that a high risk situation exists, a variety of strategies can be considered and implemented to control microbial growth.

Those outlined below are strategies which have been practically implemented; others are theoretically possible but there are many practical problems in application to overcome. The most appropriate strategy will depend on the vessel and its operating conditions and a combination of strategies is likely to be most appropriate.

5.1   Coatings

Good quality coatings provide resistance to microbially influenced corrosion as well as non-microbiological corrosion. Many coatings, however, will be attacked and degraded by microbes; these include so-called soft coatings (particularly those formulated from natural oils) and some polyurethanes. Obviously, once the coating is lost so is protection against MIC. Additionally, the environment under partially-degraded coating flakes can be particularly conducive to microbial growth and SRB activity. Glass-based coatings and epoxy coatings are generally considered to be resistant to microbial attack and some, for example coal tar epoxies, may actually have a degree of anti-microbial activity which prevents colonisation. There can be practical difficulties in retro-coating, preparing the surface and in producing a sound, adherent coating, particularly in bilges. Any defects may be the focus of oxygen gradient pitting, accelerated by aerobic microbes present, and not necessarily dependent on the intervention of SRB. In some cases it may be advisable to chemically decontaminate the steel surface before coating it. Ideally, careful consideration should be given to a suitable coating at the time of building. In large volume, high-turnover tanks, a coating may be the only cost effective measure. Coatings can be applied to high risk areas of the tank only.

5.2   Cathodic Protection

Suppressing electron flow by cathodic protection, in its simplest form by sacrificial anode, may have a limited effect against MIC but generally it is a strategy which should only be considered in combination with others, for example a good quality coating. Cathodic protection may reduce corrosion rates but the microbial populations present would continue to flourish and would become corrosive in any poorly-protected locations.

5.3   Broad Spectrum Biocide

The use of biocides for one-off decontamination when a problem has been detected has already been discussed above. Some biocides may additionally be suitable for continuous use to continuously kill or suppress bacteria (including SRB), yeasts and moulds. The effective life of biocides is finite (it may be days or weeks) and some biocide concentration monitoring may be needed to plan a biocide top-up regime. Generally, a lower concentration can be used for prevention than for decontamination. Ideally, the biocide will have good persistence, but not to the extent that it causes disposal problems. Environmental and health and safety considerations will be paramount. The factors which influence the effectiveness of one-off treatments are largely applicable to continuous treatments. Continuous biocide treatment is usually a practical measure for bilges but in non-permanent ballast tanks, where high turnover of tank contents will cause rapid loss, continuous biocide treatment is not cost effective and additionally poses disposal problems. In all cases, before disposal in discharged water, biocides may require de-activation or substantial dilution.

If custody of biocide-treated water is transferred, the receiver must be made aware that he is handling and disposing of a toxic chemical: the onus and expense of neutralising the toxic effect passes to him. Any effect of biocides on the oil/water separator must be considered.

5.4. Narrow Spectrum Biocide

The use of a narrow spectrum SRB-specific biocide for decontamination has been discussed in section 4. This biocide is also ideal for continuous treatment, and ship trials have shown it to be a cost effective option. The biocides persistence means that re-dosing is typically only necessary every one to three months.

It may be appropriate and more economical to regularly remove sludge and water to a slop tank and then target treatment of this tank. The merits of this approach will be dependent on the tank type, contents and operating conditions.

6. Conclusions

The role of microbes in accelerating corrosion in bilge and ballast tanks has been well documented. Monitoring devices are available and remedial and avoidance strategies have been developed. Although this chapter has dealt particularly with MIC in these locations, much of the information is relevant to MIC in crude oil and petroleum product carriers, to offshore facilities and to floating oil storage vessels.

7. References

1. Hill, EC and Hill, GC, 1992. *Microbiological Problems in Distillate Fuels,* Trans. I. MarE., Vol 104, pp119-130.
2. Hill, EC and Hill, GC, 1993a. *Microbes in Fuels, Lube Oils and Bilges – Recognition and Monitoring,* Workshop, Institute of Marine Engineers, London 23 Feb 1993.
3. Hill, EC and Hill, GC, 1993. *Microbial Proliferation in Bilges and its Relation to Pitting Corrosion of Hull Plate of In-shore Vessels,* Trans. I. Mar E, Vol 105, pp 175-182.
4. Hill, EC and Hill, GC, 1994. *Microbial Pitting Corrosion, Old Problems in New Places – Mechanisms, Recognition and Control,* Proceedings of the International Conference on Marine Corrosion Prevention 11-12 Oct. 1994, London, The Royal Institution of Naval Architects
5. NACE Standard TM0194-94, Standard Test Method, *Field Monitoring of Bacterial Growth in Oilfield Systems,* 1994, NACE International.
6. Pritchard, AM et al, *Critical Review of Microbially Influenced Corrosion,* AEA Technology Report AEA-TPD-51, 1994 AEA Technology.

5. Potable, Fresh and Recreational Water

WG Guthrie

Contents

1. Introduction

Water is an essential component for the growth and survival of a wide variety of microbes, some of which may be potentially harmful to man. Inevitably, therefore, wherever water is supplied for drinking, cleaning, cooking or recreational purposes, the potential for contamination exists and this can result in economic losses and may, in some cases, constitute a health risk.

Onboard systems which rely on fresh water include the potable supply for crew and passengers (including food preparation), the domestic water used for washing and cleaning purposes, the fresh water used in cooling systems for machinery and the water for recreational uses such as swimming pools and spa tubs. Uses in air conditioners and humidifiers will be dealt with separately.

2. Potable Water

Stringent statutory measures exist to control the microbiological quality of potable or drinking water. Treatment and handling requirements are set out in the relevant Merchant Shipping Notices M1214 (amended by M.1401)and M1216[1]. These apply not only to water taken from barges or shore lines but also to water produced onboard using evaporators or reverse osmosis systems. The provision of safe drinking water onboard UK registered ships is the responsibility of the ship's master and employer.

The long-established method of sterilising drinking water onboard ships and of disinfection of the tank and system is by means of chlorination. The UK Maritime and Coastguard Agency publications give recommendations on the concentration of sodium hypochlorite to be used for safe and effective treatment of the water supply. Treatment guidelines for produced or loaded water mandate the use of chlorination to achieve a residual free chlorine level of 0.2ppm at drinking water outlets. Simple colourimetric kits are available which allow easy monitoring of chlorine residuals. The effectiveness of chlorine can be reduced by the presence of organic matter in the system or if the pH of the water is outside the optimum range of 7.2 – 7.8. Other treatment regimes, such as UV light, can be considered but at present they are only permitted as an adduct to normal chlorination. Notice M.1401 accepts electro-silver ionisation for the automatic treatment of fresh water produced onboard between the production unit and the storage tanks. A minimum of 0.1ppm added silver is stipulated which ensures a residual concentration of 0.08ppm; there is no possibility of checking this onboard and an independent annual check by a competent laboratory is required. Electro-silver ionisation units for passenger ships must be approved individually.

2.1 Contamination

Extreme care must be taken to prevent any contamination of potable water, particularly where it is stored in holding tanks for any length of time.

It is recommended that storage tanks are opened up, emptied, ventilated and inspected at intervals not exceeding 12 months and thoroughly cleaned. The tanks and distribution system should be then superchlorinated according to Maritime and Coastguard Agency guidelines. The distribution system should be maintained according to manufacturer's instructions.

Contamination may occur when dirty hoses are used for filling operations. Ships' fresh water hoses should be used solely for the purpose of transferring water from shore mains supply or water barges. They should always be drained, capped and properly stowed off the deck in a safe place between use and, as a routine, disinfected every six months using super-chlorinated water at 100ppm for a contact time of one hour. Deck filling points must be protected and secured with covers. Hoses should be flushed through before use and fitted with collars to prevent their connections from coming into contact with the ground or deck. If possible, quayside potable water hydrants and filling hoses should be checked for cleanliness, leaks etc, paying particular attention to shore-side storage of hoses. Defects found should be reported to the local port authority or port health authority.

There is also a potential for microbial growth to develop in intake filters or in water-softening units if these are required to make the water more palatable. Filters must be subject to regular checks and cleaning, as should any water-softening units. Wherever these softening units are used they must be fitted upstream of any chlorination unit to ensure that contamination is effectively controlled.

Also, if the water is produced onboard either by evaporators or reverse osmosis units, these systems should be inspected regularly and maintained effectively. Where evaporators are used, operators are advised to avoid processing inshore or estuarine waters (not within a 20 mile radius) to minimise the risk of contamination with raw sewage. Suction devices must also be sited forward of, and on the other side of the ship from sanitary or bilge water discharges.

However, if these strict procedures are not adhered to then contamination may occur. This can involve the supply system being colonised by non-harmful bacteria and/or fungi which impart off-odours or taints to the water. Though unpleasant, these occurrences are not life-threatening. Remedial action would require thorough disinfection of the water system to the appropriate standard but should also involve a detailed inspection to identify whether there were any sites around the system where colonisation could have occurred. This is particularly important if there has been any repair or maintenance work done which may have inadvertently introduced dead-legs or used non-standard materials. It is now recognised that certain materials can actually encourage microbial growth by providing a substrate and nutrients. Advice and guidance on the use of appropriate fittings and materials is available from a number of sources[2,3].

2.2   Human Health

Of most concern is the potential for contamination with organisms which affect human health. Some of the more common culprits in this category and a few less common ones are listed in Table 1

Causative Micro-organism	Health Effects
Escherischia coli	Sickness
Salmonella spp	
Giardia	Diarrhoea
Amoebae	General Gastric Upset
Cryptosporidium	

Table 1. Potentially harmful microbes which may colonise potable water

The majority of these disease-producing organisms originate from human or animal waste products and are a potential hazard if potable water supplies are adulterated with faeces, urine or raw sewage. Their presence results in gastric upsets which can range from mild discomfort to serious disease which may be fatal in infants and the elderly.

Onboard microbiological monitoring of potable water is not possible because of the methods required to test for presence/absence of particular pathogenic organisms. This further underlines the importance of operating a strict checking procedure for the treatment of drinking water supplies. If contamination is suspected, then external agencies must be employed to sample and test the system.

Experience shows that the quality of fresh water can be taken for granted, particularly when water is taken at regular, frequent intervals from UK ports by a coasting vessel for example. It is generally assumed that no additional action to disinfect the water is required as 'it is the same as that I use at home'. However, some operators, as a matter of routine, arrange to have fresh water samples taken at regular intervals for bacteriological and chemical analysis using the local port health authority - this would seem a sensible approach which may help operators decide whether additional disinfection of fresh water is indeed necessary.

A maintenance log, listing all units in the system and detailing maintenance carried out, sample frequency and results, should be kept.

3.   Fresh Water Systems

Non-potable fresh water is used in applications such as recirculating cooling systems for machinery and in hot water systems for personal hygiene and general cleaning. The conditions and requirements differ for these uses and they will be dealt with separately.

3.1   Cooling Systems

In closed-circuit high-pressure cooling systems found, for example, on diesel engines, the temperatures are such that microbial contamination is rarely, if ever, a problem. However, in open recirculating systems, the temperatures are usually significantly lower and the secondary cooling circuit is normally connected to an evaporative condenser of some type which is open to the atmosphere and thus prone to external contamination. Some of the common microbes which colonise open recirculating systems and the problems they cause are listed in Table 2

Causative Micro-organism	Symptoms
Pseudomonas spp	Loss of heat exchange efficiency
Enterobacter aerogenes	
Klebsiella spp	Pipework blockage
Aspergillus niger	
Saccharomyces cerevisia	Loss of treatment chemicals
Desulphovibrio spp	Corrosion of pipework
Desulphotamaculum spp	Toxic gas production
Legionella spp	Legionnaire's disease
	Pontiac fever
Microbial endo- and exotoxins	Sick Building Syndrome

Table 2. Common contaminating organisms in recirculating cooling systems

The microbes in these recirculating systems have ideal temperature and nutrient conditions for growth. Many, particularly the Gram negative bacteria, have an affinity for surfaces and readily attach themselves to pipework and heat exchangers to form slime layers or biofilms. These biofilms create a barrier to efficient heat exchange and may in extreme cases cause blockage of pipes. An example of the extent to which heat exchangers can become grossly contaminated with microbial slimes is shown in Figure 1. The bacteria in these slime layers extract nutrients from the circulating water and in so doing can consume important treatment chemicals such as scale- and corrosion inhibitors.

Within the biofilm, oxygen gradients are produced leading to anaerobic conditions at the interface between the slime layer and the pipework.

This environment encourages the growth of specially-adapted anaerobic bacteria whose by-products are corrosive to metals. In Figure 1 it is readily apparent that anaerobic bacteria have colonised the system because of the black layer around the circumference of the pipe caused by sulphide corrosion products.

Under some conditions the cooling system may become contaminated with potentially pathogenic bacteria including *Legionella pneumophila*, the causative vector for legionnaire's disease. This problem will be dealt with in more detail in Chapter 8.

Figure 1. A heat exchanger contaminated with microbial slime

3.2   Monitoring and Treatment

All of the above factors underline the importance of careful monitoring and maintenance of these open recirculating systems. For that reason it is important to devise and implement protocols for each system which take account of the following activities:

- Regular inspection to check for slime build-up in visible parts of the system and debris in reservoirs.
- Regular monitoring of water samples for microbiological contamination.
- Use of an integrated treatment regime involving effective control of scale, corrosion and microbiological fouling.
- Minimum twice yearly planned shut-down of the system for cleaning and decontamination.
- Controlled repair and maintenance procedures to ensure that any work done on the system does not introduce problem sites, eg dead-legs.

Generally, for land based systems, the majority of the above activities would be carried out by a water treatment service company. With onboard installations, the ship's engineering staff rarely have the luxury of employing contractors to service their units. However with appropriate guidance, effective procedures can be drawn up and operated.

The visual inspections required are no more than common sense would suggest. The microbiological monitoring can be achieved by the use of a simple dip-slide. These are small nutrient agar-coated plastic paddles with an integral incubation container. The paddle is removed from the container, dipped into the cooling water at an appropriate point and returned to the container for incubation, usually at 25°C. After 24-48 hours, a semi-quantitative assessment of the microbes present is shown by the number of coloured colonies on the agar surface. Figure 2 shows the type of results which are achieved when a simple dip-slide is used to check water quality. These devices are not able to identify specific organisms but are a useful tool for monitoring the condition of a system and can identify if significant changes occur which need remedial action.

Figure 2. Dip-slides showing various degrees of contamination

For chemical treatment, the system operator has a wide choice of proprietary scale and corrosion inhibitors, the selection of the appropriate combination being a function of the materials of construction of the system and the quality of the water used. General guidance on the choice of these types of chemical will usually be provided by the manufacturer of the cooling system.

However, the choice of a suitable microbiological control programme requires more careful thought due to the range of chemicals available and their different properties and functions. The chemicals available are not necessarily appropriate for use in potable water systems.

Classification	Examples	Uses/Limitations
Oxidising	Chlorine precursors, eg. sodium hypochlorite Bromine precursors eg. Bromochlorohydantoin Hydrogen peroxide	Short term disinfection and decontamination. Can cause corrosion and can be affected by presence of organic matter
Non-oxidising	Isothiazolinones Bromonitropropanediol Glutaraldehyde Quaternary ammonium compounds Dithiocarbamates	Deliver residual activity. Not affected by presence of organic matter. Not corrosive at normal use levels. Need careful selection to optimise activity

Table 3. Biocide types and their application

Microbial control chemicals, or biocides as they are commonly termed, are divided into two main categories based on their functionality or mode of action. Some of their key active ingredients are listed in Table 3 together with their main uses and limitations. It is important to recognise that no one chemical can be viewed as a universal panacea, and the use of single biocide programmes often leads to a build-up of resistant or tolerant organisms which will eventually overwhelm the treatment and cause gross fouling of the system. The recommended approach is to adopt a regime involving two or more biocides used in combination or sequentially in a cycled programme to give a more effective outcome.

Where there is ready access to the cooling system, the chemicals may be added manually. However, with increasing concerns over exposure to potentially hazardous materials and with the level of sophistication demanded by combination and sequential dosing, it is now common practice to install automatic dosing by means of metering pumps. In this instance it is extremely important to check the performance of these pumps on a regular basis to ensure that the correct level of biocide is being added.

Care must also be taken to ensure that the biocides used are mutually compatible and also compatible with the other scale and corrosion inhibitors used in the system. This is of particular importance where quaternary ammonium type biocides are used since these chemicals react with normal anionic salts and may become inactivated but may also affect the other treatment chemical. Quaternary biocides frequently produce foam and may require the use of anti-foam agents.

If cooling water is used as the heat source for a fresh water evaporator there could be a risk of leakage of water and any chemicals it contains into the produced fresh water; the chemicals should therefore be approved for this application.

4.   Hot Water Systems

Hot fresh water is required in large quantities aboard ship for a variety of uses ranging from personal hygiene in the form of wash water and showers, to general cleaning and laundry. In general, the systems take water from storage or onboard evaporators and pass it through a calorifier unit before it is distributed through lagged piping to the point of use. The calorifiers may be heated electrically, by steam or by engine exhaust gases in some cases

With the temperatures of 50 - 60°C normally achieved in the calorifier units, it is rare for hot water systems to suffer from significant microbiological contamination problems in the form of slime build-up or corrosion. However, there have been several instances where outbreaks of legionnaire's disease have been caused by contamination of a hot water supply system. The evidence points to two possible sources for this contamination which may be present separately or together.

Firstly, the calorifier itself may become a breeding ground for these pathogenic organisms. If the unit is poorly designed or maintained, then temperature gradients can be caused where large parts of the calorifier may be consistently below the target temperature. Also, in poorly designed units, debris and rust products can build up in inaccessible areas and provide an ideal environment for microbial growth.

Secondly, the organism has been shown to colonise shower heads. This is a particular problem where long pipe runs or poor lagging mean that the temperature of the hot water at these points in the system drops to a level where the legionella bacteria can survive and multiply. Particular problems have been noted with shower units furthest from the source of hot water or in rooms/cabins which have infrequent use. Recommendations have been made that all hot and cold water supplies should be installed as, or converted to, a ring main system. This would help to prevent the formation of stagnant low-temperature zones in the supply network. These factors mean that the engineering staff aboard ship must be aware of potential problems and ensure that all hot water generating units are operating effectively and uniformly to avoid the likelihood of contamination.

Regular monitoring of onboard hot water systems for the presence of legionella bacteria is not a feasible option due to the expertise and time needed to conduct the tests. In this situation a strategy of preventive measures is advised along the lines suggested in the UK Health and Safety Executive Guidance Document, HS(G)70 (4) and reflected in the Merchant Shipping Notice M1214[1]. These call for water temperatures in the calorifier to be maintained above 60°C or for the temperature to be raised to 70°C for one hour on a regular basis to ensure that any contamination is sterilised.

In extreme cases there may be a need to treat the whole system with high levels of chlorine and this will involve flushing the system through with treated water until residual levels of up to 50ppm are measurable at the furthest outlet.

This is often referred to as 'super-chlorination'.

The general advice is to conduct a rigorous programme of inspection and monitoring covering the condition of calorifiers and pipework as well as the temperatures achieved in the calorifier and at the outlets.

There is also a requirement to ensure that shower heads and associated tubing are removed and disinfected every three months using super-chlorination. If all these measures are implemented, the risk of infection from hot water systems should be eliminated.

5.   Recreational Water

In recent years there has been a significant increase in the use of fresh water for recreational purposes onboard ship. The growth in cruising as a leisure and holiday pursuit has spawned ever larger cruise ships with up to three swimming pools and associated spa tubs and whirlpool baths. Some ships also have more exotic water features such as slides and fountains. Even the more humble merchant vessels increasingly are equipped with a small pool and/or spa tub. These recreational fresh water systems bring their own special set of problems where microbiological contamination is concerned. An excellent guide to the operational factors relating to recreational waters has been published by the Pool Water Treatment Advisory Group in the UK[5].

Some of the main organisms associated with swimming pools and spa tubs together with the problems caused are listed below in Table 4.

Causative Microbe	Symptoms
Non-specific algal growth	Discoloration of water Scum formation
Faecal and urinary tract organisms	Sickness
Giardia *Cryptosporidia*	Diarrhoea
Legionella pneumophila	Legionnaire's disease Pontiac fever
Tinea pedis	Athlete's foot
Mycobacterium marinuma	Skin granuloma
Pseudomonas aeruginosa	Folliculitis Otitis externa
Naegleria fowleri	Meningitis

Table 4. Microbes associated with swimming pools and spa tubs

5.1   Contamination

Algal contamination of swimming pools is relatively rare and is generally symp-tomatic of pools where the hydraulic flow is poor or where a pool is left untreat-ed for any length of time. Because of the need for light as part of their metabol-ic processes, algae are only problematic in outdoor pools or in covered pools where daylight can penetrate. Their presence is unsightly but does not generally constitute a health hazard. If algae do become a problem, this can be dealt with by removing the visible growth mechanically with a vacuum cleaner. If this proves unsuccessful, then the normal chlorine disinfectant levels can be increased. Failing that, a proprietary algicide based on a quaternary ammonium, polyoximinio or copper-based compound, can be added.

Swimming pools and spa tubs can be regarded almost as a special case when considering potential health hazards from microbiological contamination. Due to the large numbers of bathers usually involved, the intimate contact of the bather with the pool water and the likelihood for ingestion, swimming pools can be viewed as a high risk area if the pool water becomes contaminated. Ironically, however, the bathers themselves are the most likely contributing factors in the cycle of contamination. As can be seen from Table 4, several of the microbio-logical problems arise from waste products released accidentally or deliberately or from infected sites eg faecal matter, athlete's foot. With the intimate contact involved it can be appreciated that diseases will be transmitted readily if these organisms are not well controlled.

Health problems may also be caused if other non-transmitted organisms are allowed to colonise the pool water. For example, pathogenic strains of *Pseudomonas aeruginosa* have been implicated in a condition known as 'swim-mer's ear' or otitis externa. These organisms have also been shown to cause fol-liculitis in whirlpool/spa tubs where the temperatures are usually higher than in a normal pool and the exposure time of the bather may be significantly longer. Where there is a risk of aerosol generation, for example near the surface of spa tubs or with features such as water curtains or fountains, the operators must be aware of the potential hazards from contamination with *Legionella pneu-mophila*. The infective route for this bacterium is through inhalation of aerosol mists and, therefore, care must be taken to ensure its absence from the bathing environment.

The microbiological monitoring options are somewhat limited when it comes to pool waters onboard ship. Most of the testing requirements call for specific methodologies which could only be carried out in a microbiology laboratory. This is particularly true when testing for the presence of potentially pathogenic organisms. However, for general indications of microbiological quality it may be acceptable to use the dip-slides mentioned earlier (Figure 2). The general con-census seems to be that if the pool is well maintained in terms of cleaning, chem-ical treatment and pH control, then problems from microbial contamination should be insignificant.

5.2   Importance of Disinfection

The importance of appropriate disinfection regimes cannot be stressed enough and some of the main treatment approaches are listed in Table 5. It will be seen that the cornerstone in the pool treatment armoury is the oxidising disinfectant biocide. Traditionally this has been chlorine, or more accurately hypochlorous acid, since this is the chemical species that does the actual killing. Treatment may be in the form of chlorine gas or sodium hypochlorite. The chloroisocyanurates act as a source of available chlorine by decomposing in water to release the active disinfectant. In chlorine treated pools it is vital to maintain the pH within the range 7.2 to 7.8 to achieve optimum performance. If the pH is not controlled well, then overdosing may occur, leading to unpleasant or even dangerous conditions for the bather.

Disinfectant	Source	Target Dose	Conditions
Chlorine (hypochlorous acid)	Chlorine gas Sodium hypochlorite	1mg/l free chlorine residual	pH 7.2 - 7.8
	Chlorinated isocyanurate	2.5-5mg/L free chlorine residual	pH 7.2 - 7.8
Bromine (hypobromous acid)	Liquid bromine	2.0 - 2,5mg/L total bromine residual	pH 7.8 - 8.2
	Bromochlorohydantoin	4.0 - 6.0mg/L total bromine residual	pH 7.2 - 7.8
	Sodium bromide plus hypochlorite	2.0 - 2.5mg/L total bromine residual	pH 7.8 - 8.2
Ozone (O3)	Ozone generation (electrolytic)	0.8 - 1mg O_3/L	pH 7.2 - 7.8 normally used in conjunction with chlorination
Ultraviolet radiation	Flow through UV chamber	Not applicable	pH 7.2 - 7.8 normally used in conjunction with chlorination

Table 5. Commonly-used swimming pool disinfectants

The key control parameters to ensure safe conditions in swimming pools and spa tubs are maintaining the correct pH balance and ensuring that the correct residual levels of disinfectant are achieved. Both of these measurements can be carried out with simple test kits so that the onboard engineering staff can check and regulate the pool treatments to prevent any microbiological problems.

The choice of chemicals for pH control will generally depend on whether the disinfectant being used is alkaline or acidic. For example sodium- or calcium hypochlorite are alkaline and normally require the addition of acid for pH correction. Sodium bisulphate, hydrochloric acid and carbon dioxide are appropriate is these cases. For acidic disinfectants such as chlorine gas, liquid bromine or trichloroisocyanuric acid, sodium carbonate (soda ash) is the usual additive for pH adjustment.

Particular care should be taken with spa tubs and whirlpool baths. These devices normally run at a higher temperature than a swimming pool which will encourage more microbial growth and may also make chlorine-based disinfectants less effective. With the agitation normally involved they can generate aerosols which could be inhaled by the bathers who may spend significantly longer periods in these units than in a swimming pool. It is vitally important therefore that whirlpool and spa tubs are monitored very closely and that their treatment is closely controlled to minimise the risk of infection.

As with many of the situations highlighted in this chapter, there is no substitute for having well-thought-out procedures for the maintenance and control of the various fresh water systems used for recreational purposes. When these are implemented and operated by all the staff, then the problems normally associated with microbiological contamination will be avoided.

6. References

1. Department of Transport UK Merchant Shipping Notices M.1214 (1986), M.1216 (1986) and M.1401 (1989).
2. List of substances and products approved for use in the production of potable water from seawater or brackish water and for the treatment of swimming pool water. Committee on Chemicals and Materials of Construction for use in Public Water Supply and Swimming Pools, DoE Drinking Water Inspectorate, London
3. British Standard 6700: 1987 Specification for the design, installation, testing and maintenance of services for supplying water for domestic use within buildings and their curtilages. Available from: British Standards Institution, Sales Department.
4. HSE HS (G) 70 *The control of legionellosis including legionnaires' disease.* HMSO 1993 ISBN 0 11 882150 4
5. *Pool Water Guide: The Treatment and Quality of Swimming Pool Water,* The Pool Water Treatment Advisory Group, 1995 ISBN 0 9517007 1 5

6. Ventilation System Hygiene

A Webster and L Baxter

Contents

1.   Introduction

Ventilation system hygiene is of increasing interest both ashore and at sea. Outbreaks of legionnaire's disease and concern regarding 'sick building syndrome' have been widely publicised. These and other associated problems can have an adverse effect upon indoor air quality and the health of the ventilated space's occupants. Mould growth in the heating, ventilation and air-conditioning (HVAC) system and associated ductwork also presents problems with respect to day-to-day operation and maintenance. The reduction in air flow, which is often associated with mould physically blocking ventilation ducts or recirculation units, may present a whole range of 'knock on' problems, relating to thermal comfort, air exchange rates and a concomitant build up of pollutants.

Some research has been conducted in land-based situations over the past couple of decades but little has been done at sea. One of the few studies to determine the scale of shipboard ventilation system hygiene problems and the principal causal factors in ductwork contamination has been carried out by Lloyd's Register of Shipping through a number of investigations on cruise vessels operating in the Caribbean and the Mediterranean. The results of the study indicate that the indoor air was generally of good quality, but the HVAC systems themselves had the potential to develop contamination problems due to a number of factors, for example:

- Poor design – eg, inappropriate siting of air intakes and outlets, poor access to the air handling unit (AHU) and ductwork, no pre-filtration.
- Poor maintenance – eg, inadequate cleaning of cooling coils.
- Poor implementation of procedures and calibration of systems – eg, no log of filter changes, unknown filter specifications.
- Ignorance of potential environmental health problems – eg, microbial contamination simply regarded as dirt or dust.

The following outlines the sources and effects of microbial growth in HVAC systems and offers practical advice to engineers with respect to limiting this contamination.

2.   The Microbes

The most common microbes to infiltrate the HVAC system are fungi (particularly moulds), bacteria and viruses. The sources of these biological contaminants are shown in Table 1.

2.1   Moulds

Contamination of shipboard HVAC systems typically appears in the form of black/brown deposits which line the walls of the ductwork and accumulate in low pressure areas of the system, such as cabin reheat units. On certain ships,

Sources	Fungi & moulds	Bacteria & Viruses
People	-	2
Moist construction materials	1	1
Air refrigeration units	2	1
Ventilation system	2	1
Evaporation coolers	1	1
Humidification system	3	1
Wet cooling towers	-	1
Water leaks	2	-
Other water systems	1	1
Moist basement	2	-
Deteriorating building materials	2	-
Fabrics	-	1
Other surfaces	2	1
'House' dust	1	1
'House' plants	1	
Outdoor air	2	1
Food	1	-

Table 1. Major sources of moulds, bacteria and viruses[2]

this microbial growth has been found to completely clog ducts and reheat units on a regular basis.

Spores of moulds gain entry to ships either directly via open doors and windows or through the inlets to the HVAC system. The inlet air handling units of ships usually incorporate some form of filtration system which prevents the ingress of particulate matter. However, dirty and ill-fitting filters often allow dust and spores to pass unimpeded into the HVAC ductwork.

Most microbes require moisture, organic nutrients, and warm temperatures in order to grow. Once the moulds have gained access to the ship it is water that is the main factor affecting growth. Therefore, microbial growth occurs in those parts of the system prone to moisture incursion or condensation. Following a period of growth, spores are distributed further into the ductwork and eventually may spread throughout the ship. Once microbial growth has been established in the HVAC system it is very difficult to remove.

Many of the most prolific types of mould found to be present in HVAC ductwork, such as *Cladosporium* and *Penicillium*, are not particularly harmful to human health. However, these species may play host to other more virulent species of moulds, bacteria and even viruses which may be of much greater risk to health[1]. For instance, *Aspergillus* colonies have been found on samples containing high *Cladosporium* contamination.

2.2   Bacteria

Most environments contain a wide variety of bacteria. Within a ship, people, and particularly their skin, mouth and nose, tend to be the main source of bacteria (micrococcus, staphylococcus) and, therefore, the level of contamination is largely dependent upon the hygiene and behaviour of the occupants. Bacteria from external sources are much less of a problem due to the sterilising effect of sea water and sea spray.

Airborne bacteria cause infection by entering the respiratory system of the host and, in recent years, there has been much concern regarding the risk of legionnaires' disease aboard passenger ships. However, opinion varies as to whether *Legionella* contamination can actually occur in ships' air-conditioning systems. Nevertheless, there is a risk of *Legionella* proliferation within parts of the HVAC system, particularly in the humidifier reservoir, but of much greater concern is the risk of *Legionella* infection from spa pools and showers where warm water readily forms aerosols in the breathing zone (see Chapter 5).

2.3   Other Biological Agents

Various allergenic microbiological contaminants also affect the quality of the indoor air, but generally to a lesser degree than in land-based situations. The main sources of human-derived allergens released into the indoor environment are from dander and shed epidermal cells. Once inhaled, these allergens may trigger allergic asthma or rhinitis. Allergens can also be found inside pollen grains. Fragments of pollen may be more allergenic than the whole pollen[2] and may enter via the HVAC by passing the filtration system more easily. Dust mites feed upon skin scales and are a major problem in temperate countries as their microscopic faecal matter causes allergic reactions amongst asthma sufferers when inhaled.

2.4   Health Effects

A range of infections and allergies (Table 2) are attributed to microbes and other biological contaminants, with allergy being the most common problem. These allergies take several forms ranging from asthma, rhinitis and eczema, and the less common diseases such as allergic bronchopulmonary aspergillosis.

The extent of health problems caused by exposure to microbes onboard ships is not clear. The procedures for recording illness among passengers and crew do not allow for the collection of any meaningful data. Anecdotal evidence suggests that respiratory illness caused by HVAC contamination on passenger ships is widespread. Given that the majority of vessels have air-conditioning systems and that in most cases these systems are prone to contamination, it is likely that the symptoms similar to those experienced in 'sick' buildings, such as nausea, dizziness, mental fatigue and eye, nose and throat irritations, could be encountered.

Condition	Symptoms
Airborne contact dermatitis	inflammation of the skin
Allergy	undesirable physiological events mediated by specific allergens
Alveolitis	inflammation of the lung resulting in breathlessness
Aspergillosis	specific form of asthma
Asthma	variable airways obstruction and bronchial irritability
Conjunctivitis	irritation of the eyes (conjunctival mucosa)
Eczema	skin rash in predisposed individuals
Endotoxicosis	toxic response to bacteria
Humidifier fever	influenza like illness caused by endotoxins
Mycotoxicosis	toxic response to moulds
Otitis	inflammation of parts of the ear
Pneumonia	infection of the lung often by a specific bacterium (eg *Legionella pneumophila*)
Rhinitis	irritation of the nose (nasal mucosa), itching, sneezing, running or blocked nose
Sick building syndrome	several symptoms including: • eye, nose and throat irritation • skin erythema, dryness of mucous membranes • headache, nausea, dizziness, mental fatigue • airway infection, cough • hoarseness, wheezing
Sinusitis	irritation of the sinuses.

Table 2. Possible health effects caused by microbiological contamination of HVAC system

3. Major Sources of Contamination and Remedial Measures to Limit Risk

3.1 The HVAC System

Virtually every ship in service today employs some form of mechanical ventilation system. The word ventilation is derived from the Latin Ventus, which means the causing of air movement or wind. Today, ventilation usually refers to the provision of air changes in an enclosed space to admit fresh air in order to replace stale air. Ventilation provides oxygen, dilutes carbon dioxide and other undesirable gases and reduces odours. The extremes in climate encountered by ships and their high internal space usually means that air-conditioning is essential.

The HVAC system needs to provide air of good quality which is dust- and odour free with a minimum of noise and draught, and to maintain thermal comfort under a wide range of operating conditions.

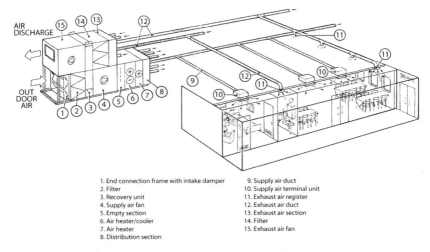

1. End connection frame with intake damper
2. Filter
3. Recovery unit
4. Supply air fan
5. Empty section
6. Air heater/cooler
7. Air heater
8. Distribution section
9. Supply air duct
10. Supply air terminal unit
11. Exhaust air register
12. Exhaust air duct
13. Exhaust air section
14. Filter
15. Exhaust air fan

Figure 1. Principal components of an HVAC system

These requirements have to be balanced against SOLAS fire safety considerations and the operator's desire for low power consumption, low equipment space and weight, low maintenance costs and simplicity in design and operation. The principal components of an HVAC system are shown in Figure 1 and are listed below:

- Air intake
- Pre-heat coil
- Filters
- Cooling coil
- Drain pans and pipework
- Duct after cooling coil
- Control sensors
- Fan chamber
- Fans
- Humidifiers
- Duct after humidifier
- Attenuators
- Fire dampers
- Supply air terminals
- Exhaust air register
- Ductwork
- Air balancing dampers
- Discharge grilles and plenum.

3.2   Limiting Contamination in Risk Locations

There are a number of ways in which HVAC systems pose a risk in terms of microbial contamination:

- They may selectively transport microbes from the outside air to the inside of the ship.
- They may collect dirt and moisture which provide a source of nutrients for microbial growth.
- They may collect large quantities of dormant microbes(spores).
- They may support the active growth of microbes.

Once established within a ship's HVAC system, microbial contamination can pose risks to the health and safety of the passengers and crew as well as increasing operating costs. However, good design, operation and maintenance will limit the risk of contamination.

Air intake

The siting and design of air intakes is a very important factor when considering the risk of contamination. The position of air intakes in relation to exhaust vents and external sources of pollution needs to be considered at the design stage. Where possible, intakes should incorporate some form of screen to prevent large items of debris, such as insects, birds and leaves from being drawn into the system. Intake louvres or screens should also be designed to prevent water penetration. Microbial growth is dependent upon nutrients in the form of dirt and water, therefore, preventing these nutrients from entering the system in the first instance will help to limit contamination as well as to reduce corrosion. There is an increasing trend among passenger ship operators to fit moisture-retaining pre-filters at the intake.

Pre-heat coil

Where present, pre-heat coils provide an ideal dirt-trap. Once again the presence of dirt and moisture may promote microbial growth. In addition, dirt on coils will have an insulating effect leading to a loss in performance. It is important therefore to keep coils clean for financial reasons as well as health concerns.

Filters

Filters are the most efficient way of preventing HVAC system contamination. On most ships, the air handling unit filter is the first line of defence and in essence these filters do a similar job to the air intake screens. In most circumstances, filters should be sufficient to remove microbial agents larger than 1µm or 2µm in diameter. As mentioned above, some ships utilise low efficiency, high arrestance

Figure 2. Filter frame showing gaps in filter material

pre-filters at the air intake which are a good way of prolonging the life of the main, moderately efficient filters. Filters not only prevent the ingress of microbes and spores but also the dirt and debris which act as nutrients. Typically, the cleanliness of the supply ductwork is directly related to the condition of the filters.

Generally, problems regarding filtration are associated with poor maintenance. Gaps or holes in filter material are the main route by which particulates, microbes and, in particular, mould spores enter the HVAC system (Figure 2).

Once established in the system they are very difficult to remove and therefore every effort should be made to prevent their entry at the earliest stage. Where filters are fitted correctly they should be cleaned and replaced on a regular basis as added resistance will reduce airflow and system performance.

There are no statutory requirements with respect to filter material in marine applications. European and American guidelines have been published which aim to improve air quality and prevent microbial contamination in buildings. The Commission of the European Communities recommend '...air filters of at least EU4 quality should be placed at the intake. A second filter better than EU7 should be located on the supply side of the air handling unit behind a fan or any aerosol producing device...' Further necessary filters, eg high efficiency particulate air filters should be installed close to the room[4,3]. The American Conference of Governmental Industrial Hygienists recommends filters with a dust spot efficiency of 50-70% to remove the majority of particulate matter and microbial agents from the supply air stream[1]. They also suggest that low-efficiency high-arrestance pre-filters should be fitted to improve the performance of the air handling unit filters.

Cooling coil

Cooling coils are usually present on ships, and display similar problems to pre-heat coils. The condensation which occurs on cooling coils readily attracts dirt and, unless treated, damp cooling coils can play host to slime moulds and other microbes.

Drain pans and pipework

Drain pans and the associated pipework should present few problems unless the drains become blocked. If water is allowed to collect and become stagnant, then microbial contamination may occur relatively quickly. This can be a potential source of *Legionella spp.* proliferation. However, on many ships there is poor access to drain pans, and cleaning is virtually impossible. Corrosion is also often a problem associated with drain pans and pipes. Specification of 'corrosion proof' fittings reduces the need for regular maintenance.

Fans and fan chamber

Dust and dirt accumulation in the fan chamber and on the fan blades is a common problem (Figure 3). Fan chambers are often relatively damp due to the

Figure 3. Heavily contaminated fan blades

proximity of the cooling coils and the air intakes. Therefore, if microbial growth can become established in the fan chamber, the potential for contaminating the ductwork is great. In addition, dirt on impeller blades and the scroll surface can have a great effect upon airflow performance.

Humidifiers

Humidifiers and air washers are almost always contaminated microbially and therefore need continuous attention to prevent problems. The water reservoirs associated with humidification systems provide an ideal environment for microbial growth. Water is recirculated, allowing sediment to be formed. This sediment becomes a perfect environment for microbes such as bacteria, fungi, algae and amoeba[3]. If this contaminated water is disturbed in any way, the risk of droplets, or an aerosol, being formed and being distributed into the system is great, unless means are provided to prevent this, such as the installation of moisture eliminators. Humidifiers should be cleaned and disinfected on a regular basis to help minimise and control the proliferation of micro-organisms, especially, *Legionella pneumophila*[5].

Attenuators

Sound attenuators and lined ductwork may provide a suitable media for microbial proliferation, especially when the attenuation material breaks down. Attenuator and ductwork liner material, usually made of glass fibre, can collect dirt and dust and retain moisture, allowing bacterial and fungal growth to occur[6]. Maintenance procedures should ensure that the material is sealed at all times to avoid this problem.

Fire dampers /air balancing dampers

Accumulated dirt and microbial contamination can potentially cause corrosion and damper mechanisms to stick. Dampers should always be checked following ductwork cleaning to ensure that dirt has not been forced into the mechanism.

Supply grilles

Supply grilles are usually the point at which passengers notice microbial contamination in the form of spores and debris emanating from the system. It has been suggested that further filters should be placed in the recirculating air duct, in the air-conditioning unit, between the last air handling section and the fan, and behind the supply grille. This procedure has been implemented on several ships and appears to present more problems than it is intended to solve. These additional filters allow microbes to collect on the filter material where, in the presence of moisture and other debris, they are able to reproduce. The source of the contamination is effectively trapped in the system. When supply airflow is increased, for whatever reason, some of the deposited material is liable to escape, resulting in passenger exposure. If a system is so grossly contaminated that such filtration is required then the filters should be cleaned and/or replaced on a regular basis determined by the rate of soiling. Such problems emphasise the need for regular inspection and monitoring to ensure that remedial measures are effective.

Extract grilles

Some extract grilles get very dirty and can become almost blocked with mould and debris, whereas others can be very clean, depending upon the location of the grille. In certain areas of high public use, such as, smoking areas or heavily-used passageways, whereby re-suspension of settled dust can occur, considerable amounts of dust can be found on the grilles. In these particular locations a routine cleaning regime should be adopted to ensure the grilles are kept clean for aesthetic, as well as hygienic and operational purposes.

Ductwork

Unless a management programme of inspection, cleaning and disinfection is in place to maintain acceptable indoor air quality and thermal comfort, supply ductwork is likely to be become contaminated with dirt and debris and, therefore, potentially with microbes. Contamination can be quite severe in some locations. Contamination of exhaust and recirculation ductwork is more prevalent and is typically associated with damp conditions, such as those close to bathrooms (Figure 4) and appears to be a greater problem on those vessels with a

Figure 4. Mould growth extracted from bathroom exhaust plenum

high degree of recirculation of cabin air. Contamination in these conditions can get extremely heavy and has been known to completely block the ducts.

Contributing factors include the following:

- A regular supply of water vapour from the shower which allows moulds to grow.
- Ducts are often constructed of flexible corrugated plastic material which connects the outlet to the main exhaust duct. The corrugations of this material provide a low velocity 'boundary layer' on which dirt and microbes can settle, adhere and become established.
- Metal ducts may also provide a suitable habitat especially where joints, elbows, screws and rivets provide low pressure areas for dirt and spores to settle.

Modern ships with good supply filtration appear to be equally prone to exhaust ductwork contamination which suggests that open doors and passengers' clothing are the main source. The diversity of species of microbes found in exhaust ducts is much greater than in samples from supply ducts.

It is recommended that, where possible, bathroom exhaust ductwork is constructed from material such as galvanised steel to limit the collection of dirt and spores close to a principal source of moisture. In addition, the exhaust plenum in cabin bathrooms should be cleaned on a regular basis.

Discharge grilles and plenum

The location of discharge grilles is very important as they are liable to be a major source of indoor air contamination. They should be positioned at least 10m away from the air intake to avoid contaminated air being drawn back into the system, and away from public areas.

3.3   Ductwork Cleaning

Where microbial contamination does occur then it may become necessary to clean the HVAC system. Cleaning services are a growing industry world-wide. However, cleaning ductwork on ships presents very specific problems due to limited access both in physical terms and relating to the 'window of opportunity' as the length of time between refits increases. There are only a few HVAC cleaning companies world-wide with the resources and equipment required to clean an entire ship HVAC system during a refit. However, there is a growing number of methods and procedures which have recently enabled effective HVAC ductwork cleaning on board ships. These include:

- Manual vacuuming.
- Mechanical brushing.
- Hand wiping (cleaning agents).

- High volume air blast.
- Air jetting devices.
- Ultrasound.
- Steam cleaning.
- Encapsulation.
- Chemical sprays.

Differences generally relate to the method of dislodging dirt from the duct-work. For example, techniques such as air jetting, ultrasound and mechanical brushing are all intended to dislodge the material in order to allow its removal using industrial vacuum cleaners.

Ductwork-cleaning companies generally assess ductwork against five levels of cleanliness from level one – which is clinically clean - to level five - which is cosmetically clean. However, the most practical criterion for assessing cleanliness and the need for ductwork cleaning is to determine whether the duct walls are still visible. If dirt and contamination completely obscure the duct wall then it is likely that the rough surface area will help to increase the deposition of further dirt. In this case cleaning should be recommended.

Physical cleaning methods are generally regarded as the most effective. However, problems associated with access to ductwork and the complexity of ship HVAC design means that, often, chemical cleaning and/or disinfection is the only method available. Chemical biocides are often advocated to kill moulds and other microbial growth from HVAC systems, and extensive claims are often made regarding the residual properties of these biocides. Investigations into the use of biocides for cleaning ductwork have revealed severe shortcomings with respect to their efficacy. In general terms the use of biocides for ductwork disinfection is not advocated; however, where biocides are used, two chemically-dissimilar compounds should be used to avoid problems with microbial resistance.

4.   Inspection

Inspection is an essential aspect of any HVAC maintenance programme. There are three levels of inspection which should help to provide protection against contamination and potential ill health or discomfort amongst passengers and crew.

4.1   Regular Maintenance Inspections Undertaken by the Crew

These should be undertaken on a regular basis at intervals dependent upon the size of the vessel. For instance, on a large cruise vessel inspections may be undertaken on a daily basis as part of a rolling programme of maintenance. Each component should be checked at least monthly. Crew should be made aware of what to look for. Such inspections should include:
- Air handling units (filters, chambers drip pans, fan blades).

- Terminal units (plenums and grilles in cabins).
- Ductwork (where access permits) (indicative points in the system should be identified and access provided).
- A reporting mechanism needs to be in place which allows the findings of inspections to be monitored and actions taken to be recorded (eg biocide treatment, filter replacement).

4.2 Annual/Bi-annual Surveys

A more thorough investigation should be undertaken, preferably by an independent third party, to identify any shortcomings in the ship's HVAC maintenance regime. Inspections by ductwork-cleaning companies may recommend extensive cleaning work which can be extremely costly. Independent assessments will involve similar inspections to regular maintenance inspections. A more detailed examination of the ductwork may be required which may necessitate access panels to be cut and the use of specialist equipment such as borescopes or micro-camera systems. A recommendation based on the findings should consider the most appropriate actions, given the evidence of contamination.

An indication of filter efficiency and the level of contamination within the HVAC system can be gained by sampling the air for viable microbes in passenger spaces. Hidden microbiological contamination (eg mould growth in ductwork) can be assessed by sampling for airborne microbes and spores in passenger and crew spaces. Where the mould colony counts from inside the ship are more than three times greater than outside, further investigations should be undertaken as this suggests that the contamination is present within the ship. High counts both inside and outside the ship suggest that the filters are poorly maintained and that spores are passing through the HVAC system unimpeded.

4.3 Pre-Cleaning Surveys

Where regular maintainance or third-party inspections identify significant microbial contamination then a pre-cleaning survey should be undertaken. This is typically carried out by the cleaning contractor prior to cleaning works in order to target the ducts which need cleaning. Often such contractors will use sophisticated methods (remote video cameras, etc) to show the degree of contamination prior to cleaning. They can also use the same methods to demonstrate the efficacy of their method by photographing ducts following cleaning.

It must be noted that, when carrying out surveys, many companies will use microbiological counts from wipe samples (swabs) taken from duct walls to indicate ductwork cleanliness. Such samples do not give a representative sample in relation to the quality of the air supplied by the duct and therefore should be dismissed when considering HVAC contamination. Generally, visual inspection based upon the degree of soiling should be sufficient. If a surface is completely covered by dust and dirt then it should be cleaned.

5.   Standards

5.1   Legislation

The following legislation refers to land-based situations only, but in the absence of sufficient legislation pertaining to shipboard ventilation hygiene, it can usefully be applied to ships.

Regulations addressing ductwork contamination exist in a number of countries. In Sweden, the Boverkert is the government organisation which monitors the mechanical function of ventilation systems. Offices are inspected every three years, schools and day-care centres every two years. They specify that ducts should be cleaned if dust accumulation exceeds $1g/m^2$. Guidelines recommend that regular surveys are undertaken which should include:

- Photography showing the condition of the system, including borescope photographs of ductwork.
- Dust and microbiological analysis at regular points throughout the system.
- Recommendations for improvements.

In Denmark, it is the Bygge Og Bolig Stirelsen which offers official guidance on duct cleaning. Survey guidelines are similar to those in Sweden. In Japan, Building Maintenance Law requires that buildings over $3000m^2$ should be tested and cleaned on a regular basis. The US Occupational Safety and Health Administration has also proposed rules governing indoor air quality and contractors removing microbiologically contaminated duct insulation are protected by specific laws[6]. In 1997, a European Pre-Standard[7] was released which deals with the issue of ductwork cleaning access on existing systems, and at design stage, which will enable HVAC systems to be cleaned more easily.

The European Community's 'Workplace Directive' is implemented in the UK by the Workplace (Health, Safety and Welfare) Regulations 1992[8] which sets various requirements for the provision and maintenance of workplaces. The Health and Safety Commission's Approved Code of Practice states that 'mechanical ventilation systems (including air-conditioning systems) should be regularly and properly cleaned, tested and maintained to ensure that they are kept clean and free from anything which may contaminate the air'. This was applicable from 1 January 1996 for all workplaces. To assist in compliance with the Workplace (Health, Safety and Welfare) Regulations, the Heating and Ventilating Contractor's Association (HVCA) published a Guide to Good Practice - Cleanliness of Ventilation Systems[9] in May 1998. It covers aspects from design and access to the internal surfaces of the HVAC system and system testing (inspection and monitoring) to cleaning methods and verification of cleanliness. It also takes into consideration kitchen extract systems, which are deemed a fire hazard as well as presenting hygiene and odour problems, due to the accumulation of fat and grease deposits in the body of the ductwork. In summary, the document establishes a level of particulate cleanliness verification and gives an indication of when it is appropriate to clean systems in use. Although it does not cover microbiological contamination of HVAC systems in great detail, there is world-wide research currently being undertaken which will attempt to define acceptable and unacceptable levels of microbiological colonisation.[9]

Safety and environmental inspectors should look for a maintenance routine that includes regular inspection of all ventilation systems of enclosed spaces. A logbook containing results of inspections and any cleaning operations should always be available for inspection. Such management systems are now in place on some ships.

5.2   Guidelines

The US Environmental Protection Agency (EPA) and the National Institute for Occupational Safety and Health (NIOSH) have published a guide entitled Building Air Quality[10] which aims to assist building managers to properly maintain HVAC systems. Much of the guidance provided in this publication is applicable to ship HVAC systems and provides a useful basis for air quality investigations and proper maintenance. The key international guidance relating to indoor air quality are those of ASHRAE relating to ventilation for acceptable indoor air quality[11].

5.3   Ship-Specific Requirements and Standards

There is an international standard[12] on air-conditioning and ventilation design in the passenger accommodation of ships. This standard specifies design conditions for temperature, humidity and outdoor air quantity as well as methods of calculation for the air-conditioning and ventilation of accommodation spaces. However, no account is taken of the need for inspection, cleaning and repair of HVAC systems.

The US Public Health Service's Vessel Sanitation Programme has issued shipbuilding construction specifications for passenger vessels destined to call on US ports[13] which specifies ventilation system requirements. The recommendations made with respect to air supply and exhaust systems are general in nature and aimed at addressing the major design problems currently encountered. For example they state that:
- Fan rooms shall be designed and located for accessibility to conduct periodic inspections and changing of air intake filters.
- Fan rooms shall be maintained free from accumulations of water.
- Fan rooms shall be located so that any ventilation or processed exhaust air may not be drawn back into the ship spaces.
- All cabin air diffusers shall be designed for easy removal and allow for easy access for cleaning.
- All air supply trunks shall have access panels to allow periodic inspection and cleaning.

These recommendations are useful but they do not present design specifications. Currently, there are no recommendations to include existing ships participating in the Vessel Sanitation Programme to ensure that regular HVAC inspection and cleaning are being carried out[14].

The UK Maritime and Coastguard Agency published a Marine Guidance Note MGN 38 (M+F) in April 1998[15] *'Contamination of Ships' Air-conditioning Systems by Legionella Bacteria'* to replace MSN No. 1215. This warns against the risk of legionnaire's disease, identifies main danger areas and gives examples of counter measures. It was amended in August 1998 with regard to adiabatic spray-type humidifiers.

6.   Concluding Remarks

The HVAC systems of some ships, particularly older vessels, are prone to microbial contamination. A lack of awareness of the potential problems at the design stage and subsequent lack of maintenance can exacerbate the problem by allowing moulds to infiltrate the HVAC system where they are able to collect and grow. These microbes not only pose an immediate risk of respiratory illness and allergic responses, but they are also a nuisance with respect to the general maintenance and cleaning of vessels and hence are often a cause of complaint.

Due to the complexities of shipboard HVAC design, it is proper filtration, regular inspections and monitoring, and targeted ductwork cleaning that offer the best methods of keeping the system clean and preventing the accumulation of microbial contamination. However, where contamination has occurred, cleaning of the ductwork will need to be undertaken. Physical methods such as manual vacuuming and mechanical brushing are generally deemed superior, but lack of access to the ductwork means that less-effective chemical cleaning methods are sometimes employed.

Microbial contamination and ventilation system hygiene should be an important issue for ship operators and one which needs to be addressed if shipboard indoor air quality and passenger comfort is to be maintained or improved. Cost effective improvements can be made to the design and operation of ventilation systems which reduce contamination and improve air quality. The main considerations are to prevent the ingress of water into the supply and exhaust systems and to make provision for inspection and cleaning of ductwork throughout the life of the ship.

7.   References

1.   *Bioaerosols: Assessment and Control,* Janet Macker (Ed), American Conference of Governmental Industrial Hygienists, Cincinnati, Bioaerosols Committee, 1999

2   *Indoor Air Quality: Biological Contaminants.* WHO Regional Publications, European Series No. 33, World Health Organization, Copenhagen (1988)

3.   CEC *Biological Particles in Indoor Environments. Indoor Air Quality and its impact on Man.* Report No.12. Commission of the European Communities, Luxembourg, (1993)

4.   CEC *Guidelines for Ventilation Requirements in Buildings. Indoor Air Quality and its impact on Man.* Report No.11 Commission of the European Communities, Luxembourg, (1992)

5.   HSE *The control of Legionellosis including legionnaires' disease.* HS(G)70 Health and Safety Executive, (1993)

6   Loyd S. *Ventilation System Hygiene* — a review, 8th Edition. The Building Services Research and Information Association, Bracknell (1996)

7.   CEN ENV 12 097 Ventilation for buildings — ductwork — requirements for ductwork components to facilitate maintenance of ductwork systems European Committee For Standardisation (1997)

8.   HMSO Workplace (Health, Safety and Welfare) Regulations, HMSO (1992)

9.   HVCA TR/17 *Guide to Good Practice — Cleanliness of Ventilation Systems.* Heating and Ventilating Contractor's Association (1998)

10.   EPA/NIOSH *Building Air Quality - A Guide for Building Managers.* US Environmental Protection Agency and National Institute for Occupational Safety and Health, Washington DC, (1991)

11.   ASHRAE *ASHRAE Standard — Ventilation for Acceptable Indoor Air Quality.* ASHRAE 62-2001. American Society of Heating, Refrigerating and Air-conditioning Engineers., Atlanta, USA. (2001)

12.   ISO 7547 *Air-conditioning and ventilation of accommodation spaces on board ships — Design conditions and basis of calculations.* International Organization for Standardization (1985)

13.   USDHHS *Recommended Shipbuilding Construction Guidelines for Cruise Vessels Destined to Call on US Ports.* US Department of Health and Human Services, Atlanta (Revised, August 2001)

14.   USDHHS *Vessel Sanitation Program Operations Manual.* US Department of Health and Human Services, Atlanta (1989)

15.   UK MCA MGN 38 (M+F) *Contamination of Ships' Air-conditioning Systems by Legionella Bacteria.* April 1998 (amended August 1998).

7.   Acknowledgements

The authors would like to thank Kevin Lavender for his input into this Chapter and his involvement in Lloyd's Register of Shipping's initial research, and Dr Gillian Reynolds, for her valuable contribution in reviewing the Chapter. While both authors have since left LR, Dr Reynolds continues to progress LR's environmental focus as Principal Environmental Specialist. The authors also thank Stephen Loyd of The Building Services Research and Information Association (BSRIA) for his invaluable advice and ABB Flakt Marine in giving permission to use the schematic shown in Figure 1. which was reproduced from ABB Flakt Marine Catalogue 1995/1996 Section P 34, page 2.

7. Food Storage and Preparation

J Colligan

Contents

1. Introduction

The principles of storage and hygienic preparation of food whether ashore or afloat are the same, but the application of these principles at sea and on land may be different. This chapter is presented as a practical approach to the relevant parts of respective Merchant Shipping and Food Hygiene Regulations, taking into account various shipboard conditions.

Passenger ferry and cruise ship operators are likely to have a formal food safety policy consistent with the catering industry in general. It is therefore reasonable to expect passenger operators to have formal management controls to closely monitor shipboard catering. On the other hand, it would be unreasonable to expect small coasting vessels with six or seven crew to operate a formal food safety policy, although that does not, however, exempt such vessels from maintaining a safe catering standard consistent with the service of proper meals.

The most serious risk to food safety is bacterial contamination from sources such as food handlers, flies, rats, mice, cockroaches and other insects, refuse and waste food (Figure 1). Flies and cockroaches present a serious hazard because of

Figure 1. Waste food and refuse are sources of contamination

their feeding habits and the sites they visit. Flies defecate and vomit previous meals back on to the food as they feed! Rats and mice commonly excrete organisms such as salmonellae. Contamination of food may occur from droppings, urine, hairs and gnawing. Food suspected of being contaminated by rodents must be destroyed. Raw food, particularly red meat and poultry often carries harmful bacteria but these will normally be killed off in the cooking process. Contamination usually occurs because of ignorance, inadequate space, poor design or because of food handlers taking short cuts.

Bacteria prefer warm, moist environments and if food is incorrectly stored and insufficient care is taken during its preparation, harmful bacteria will multiply rapidly. Even if food is stored and cooked properly, it can still be cross-contaminated with bacteria from raw food if, for instance, the same utensils or surfaces are used to prepare both. Food contaminated with harmful bacteria looks, tastes and smells completely normal.

The foods most commonly implicated in food poisoning cases, 'high risk foods', can be defined as cooked foods or products not requiring further processing such as cooked meat and poultry, meat products, gravy and stock, milk, cream, eggs and egg products. High risk foods can be confused with raw foods which, although containing harmful bacteria, are not high risk as they would normally be cooked and not eaten in this raw state; they are therefore described as being a potential source of harmful bacteria. A steak cooked 'rare' is safe, however, because bacteria are only present on the surface of healthy meat. On the other hand, hamburgers and other products made from minced meat require thorough cooking as any harmful bacteria which were present on the surface have been distributed throughout the mass of the meat.

No catering environment can operate without harmful bacteria being present at some time; however, small numbers of most types of bacteria do not cause illness. The storage, preparation and serving of food should therefore only be carried out where the conditions are such that bacteria are denied favourable conditions for growth and the food is not exposed to the contamination risk.

Apart from the obvious requirement for cleanliness and strict personal hygiene, there are two fundamental rules to be followed to prevent contamination and the multiplication of bacteria to harmful levels, namely temperature control and segregation of raw and cooked foods. As the design of food storage and preparation areas establishes the method of operation and the structure determines how easily it can be cleaned, it is therefore essential that the design and structure of all catering spaces accords with a food safety policy.

2. Design and Structure of Galley and Food Handling and Store Rooms

Galleys should be of sufficient size to permit safe and efficient food hygiene practices taking account of the number of meals to be prepared. The size and layout of galleys in modern cargo ships with a small number of crew, may be very similar to any compact domestic fitted kitchen. Galleys and handling rooms on vessels catering for a substantial number of persons should be designed to allow a continuous workflow to progress in a uniform direction from raw materials to finished product, to eliminate the risk of cross-contamination. Dirty (pre-cook) and clean (post-cook) processes should be separated. Other risk factors such as the distances travelled by raw materials, utensils, food containers, waste food etc and personnel should be minimised. No food rooms should connect directly with any sanitary accommodation.

2.1   Food Rooms

All food rooms should be constructed of durable material that can be easily cleaned, such materials to be smooth and non-flaking, impervious and light in colour. Bulkhead joints with deckhead and deck should be coved, as corners are difficult to access to clean. Decks require to be anti-slip and, if appropriate, suitably inclined for liquids to drain to scuppers. Drainage systems should be of sufficient fall and diameter to remove the contents effectively at all times with the minimum risk of backflow. Impervious materials such as stainless steel can create condensation, particularly in the galley deckhead, therefore sufficient suitable ventilation is essential to prevent this source of contamination. An acceptable working environment in terms of temperature and humidity must be provided with additional exhaust fans to draw off fumes and steam from cooking appliances. All areas should be adequately lit, making the best use of natural lighting where possible.

Furniture and fittings in the galley should be impervious and non-corrodible. The bottoms of all fitted furniture should be either flush with the deck or fitted high enough to enable the deck space beneath to be easily cleaned. Preparation equipment should be regularly inspected and defective items replaced or repaired.

2.2   Galley

Every galley should be provided with hot and cold water, and suitable and sufficient facilities for the efficient washing of food, equipment, cutlery, crockery etc. Mechanical dishwashers provide a very safe cleaning method.

A wash-hand basin(s) with hot and cold water, soap, nailbrush and drying facilities should also be provided to secure personal cleanliness. On small vessels, separate handwashing facilities will normally be available close to the galley.

All equipment, working surfaces and other utensils should be designed and constructed to minimise harbourage of soils, bacteria or pests and to enable them to be thoroughly cleaned and disinfected. Food-grade stainless steel is appropriate for most equipment. The use of different colours as a code to ensure equipment such as knives and chopping boards used for raw food is not used for high-risk food is recommended.

Toilets with wash-hand basins should be situated near to, but separate from, the galley. Prominent signs about washing hands should be displayed.

Suitable and sufficient containers, constructed of impervious materials with close fitting lids or covers should be provided in the galley for food waste.

2.3   Storage Facilities

Every ship should be provided with storage facilities for fresh fruit, vegetables and dry goods, in addition to the requirements for chilled and deep frozen products. All such stores should be so located as to be easily accessible from the galley and their contents protected from contamination and deterioration from exposure to weather and ship operations.

Shelves or racks should be large enough to take all stores, leaving the deck clear for access and handling. Sufficient space for cleaning should be allowed between lower shelves or racks and the deck. In addition to the careful selection of fittings suitable for the reception of the stored foods, the dimensions for the individual storage areas should be adequate for the number of crew and, if appropriate, passengers for the duration of the voyage. The term adequate can only be defined by the practical measurement of all food items and can be scaled for variations in requirements for numbers and/or time.

Failure to ensure satisfactory stowage conditions (Figure 2), temperature, humidity, air circulation, and the integrity of packaging can result in problems of unfit or spoiled food and at the very least result in a considerable reduction in shelf-life.

Figure 2. Unsatisfactory stowage can result in unfit food

All refrigeration units, chill rooms and deep freeze storage compartments should be fitted with thermometers so positioned that they accurately reflect the temperature in the food compartment and may be easily read. Recordings should be checked regularly for accuracy by using another thermometer placed in the unit, or using a probe thermometer. Chest freezers and domestic type refrigerators are unlikely to display operating temperatures and, therefore, thermometers should be kept in the units.

3. Temperature Control

It should be noted that recent legislation for England and Wales requires holding chilled food at a temperature below 8°C. It is, however, generally the case that operators maintain temperatures below 5°C to allow a few degrees margin of error and, therefore, that standard is used throughout this chapter.

Pathogenic bacteria thrive in warm conditions, therefore to prevent their growth it is essential to keep food either very hot (above 63°C) or very cold (below 5°C). Accordingly, food should not be left in the danger zone (5°C-63°C) for longer than is absolutely necessary. As a general rule, a single period of not more than two hours at ambient temperatures should be imposed for preparing and keeping food.

Thorough cooking to a centre temperature of 74°C is usually required and it would be reasonable to expect the chef on a passenger vessel to use a probe thermometer to check the temperature.

Cooling of food, particularly joints of meat, is likely to be a potential health risk as food should be cooled below 10°C in less than 1.5 hours. The process can be hastened if large joints are cut up before cooling. Whenever possible, cooling cabinets should be provided. A fan-assisted chill room may be considered in this context so long as it is only used for this process. A cool pantry or an area segregated from raw food would also be acceptable. Hot food should never be placed directly into a refrigerator as this would raise the temperature of food already being stored. Specialised blast chilling equipment is available but would only normally be supplied on busy passenger ferries, for example.

Similarly, rapid-thaw cabinets are available to defrost food but unless substantial defrosting is required on a regular basis, it could be difficult to justify the expense. Defrosting in a refrigerator is generally too slow a process. Controlled thawing of raw meat and poultry should take place in a cool area entirely separate from other foods which may be exposed to risk of contamination from thawed liquid. This area must never be used for cooked food which is cooling prior to refrigeration. Thawing can also be achieved using clean, cold running water. In some cases, thawing can be achieved using a microwave oven although extreme care is necessary because of the risk of uneven heating resulting in some parts of the food cooking before the whole mass of food is thawed. This method should be strictly limited to defrosting very small portions of food and manufacturer's instructions should be closely followed. The food handling room within the cold stores area is acceptable, provided that the area is clean and the food is covered and stored in a container. Food should be prevented from sitting in the thaw liquid by placing it on grids either above trays on a shelf or in a container. Defrosting large quantities of meat should be carried out in a cool larder at 10°C to 15°C. The large sink or bucket of cold water method, frequently observed, should be discouraged, particularly if the bucket is located in the galley. At the end of the process, flesh should be pliable with no ice crystals present in the body cavity of poultry, for example.

Chill cabinets, cold rooms and refrigerators should not exceed 5°C, and deep freeze units and cabinets should be -18°C or below. Some older systems may be unable to reach -18°C, in which case a few degrees tolerance has to be accepted. As a guide, frozen food can be safely stored at -12°C for one month only. The presence of ice usually indicates fluctuating temperatures. High humidities and fluctuating temperatures (above -10°C) accelerate mould and spoilage bacterial growth causing souring and rancidity of meat, (Figure 3). Food should never be

Figure 3. Mould growth in a food handling room

stored in front of the cooling unit as this restricts the circulation of air. Suitable packaging is essential to avoid loss of moisture from the surface of food causing freezer burn. Regular maintenance of refrigeration equipment, defrosting and checks on the correct functioning of thermometers should be carried out as a routine by shipboard personnel.

Where cold or hot buffet meals are served, suitable temperature-controlled holding units should be provided. The units should be capable of controlling temperatures below 5°C or above 63°C. Appliances such as bain-maries and hot-presses are designed for storing hot food which has been thoroughly cooked. They must not be used for warming up cold food as storage temperatures cannot be relied on to destroy harmful bacteria.

Dry food stores should be dry, and cool, about 10-15°C and well ventilated.

4.   Segregation of Raw and Cooked Products

Raw food must always be kept apart from cooked food or milk, for example, that requires no further treatment before consumption. Separate refrigerators are preferred although if in the same unit, the raw food must always be placed at the bottom to avoid drips contaminating ready-prepared food. Food should also be covered to prevent cross-contamination, drying out and absorption of odour.

Separate work surfaces, chopping boards and utensils should be set aside for the preparation of raw meat and must not be used for the preparation of foods which will be eaten without further cooking. Even in a small galley, separate chopping boards and utensils should be used rather than relying on disinfecting areas and utensils between respective operations. Colour coding is an established way of ensuring separation between the two activities.

Open cans of part-used food should not be stored in a refrigerator. Unused contents should be emptied into a suitable container such as a covered plastic bowl.

5.   Dry Goods

Accommodation used for the storage of dried and canned foods should be well lit, dry, cool, ventilated, vermin-proof and kept clean and tidy. Food should be stored off the deck. Cased goods may be stored on duckboards. Store bins with covers should be used for pulses, flour etc. Bags of flour should be stored in tightly sealed bins or silos rather than leaving them in bags. If they have to be stacked they should be cross-stacked on duckboards to permit maximum circulation of air. Provisions should not be stored against exposed steel bulkheads.

6.   Fruit and Vegetables

Due to the nature of these goods, it is essential that the packing procedure is such that protection is given to the soft items. Ideally, fruit and vegetables with one or two exceptions should be kept refrigerated. Lack of separate refrigerators usually results in ships storing fruit and vegetables in cool rooms without undue problems. Care must be taken to avoid warm moist conditions and condensation which will encourage bacterial spoilage and mould growth. Fruit and vegetables should be examined regularly and mouldy items removed to avoid rapid mould spread.

7.   Stock Rotation

Satisfactory rotation of stock to ensure that older food is used first is essential to avoid spoilage. Date codings on most foods should prevent out-of-date stock being used. Rotation should also help maintain correct stock levels. Where possible, stores should be moved to the front of shelves or, in large accommodation, to the front of the storeroom, before new stores are taken on board. Stock which remains undisturbed for long periods of time will encourage rodent and insect infestation. On passenger ships there may be a local colour coding system used to enforce strict stock control. Daily checks should be made on short-life perishable food such as fresh fruit and vegetables. The rule is 'first in, first out'.

8.   Personal Hygiene — Food Handlers

Most people carry some type of food poisoning organism at one time or another. Food handlers have a responsibility, therefore, to observe high standards of personal cleanliness to ensure that they do not contaminate food.

Food handlers should look clean and tidy and wear appropriate protective clothing, to protect the food not the individual. They should be fit for work, for example not suffering from a cold, sore throat, skin infection, diarrhoea etc. Hands should be washed regularly but particularly after visiting the WC, always on entering the galley, before handling any food or equipment and in between handling raw and cooked food.

Direct contact between hands and food should be avoided so far as possible, using tongs for example. Protective gloves may be worn but they can give a false sense of security.

Cuts, spots, sores etc should be completely covered by coloured (blue or green) waterproof dressings. The bright colour would be clearly visible if the dressing accidentally got into the food.

As the mouth is likely to harbour germs, they should not eat sweets, chew gum, taste food with the finger or an unwashed spoon. Similarly, there should be no smoking in food rooms or whilst handling open food. Although the obvious contaminant would be cigarette ash or smoke, people touching their lips whilst smoking may transfer bacteria to food. Of course, cigarette ends contaminated with saliva should not be placed on working surfaces.

9.   Refuse Storage and Disposal

Garbage must not accumulate in food areas. Food waste in the galley should normally be stored in covered containers and must be regularly removed to a designated storage area outside the galley. Some discretion may be allowed to use open black bags in a very busy galley, during peak periods for example, so long as bags are not over-filled and they are removed frequently to prevent contamination. It is unacceptable, however, to allow an accumulation of open waste to remain uncovered as a matter of routine.

Disposal depends on the scale of operation. For example, passenger ships may use waste disposal units to grind and discharge waste food in designated areas or waste can be compacted and stored in a cool holding area until it can be landed. A similar grinding process can be used to dispose of glass refuse. Small coasting vessels would only require a designated area, such as a skip or bucket probably on the deck, convenient to the galley to hold a few plastic bags which would be landed at the next port. Any system is acceptable so long as areas are clean and tidy and containers sealed so as not to attract vermin.

10. Cleaning and Disinfection

All articles that come into contact with food should be thoroughly washed, rinsed and disinfected before use. Articles include trays, knives, cutting boards, food preparation machinery and work tops. There are various proprietary chemical disinfectants and their choice depends on the amount and type of soiling, water hardness etc. Hypochlorite is generally considered to be the best disinfectant for general catering and it must be mixed to the concentration recommended on the pack. Disinfectants should be freshly mixed for each job and thrown away when the job is finished or within an hour of mixing and use. Unused diluted disinfectants should be thrown away after 24 hours as they break down on standing. NEVER add more disinfectant to a used or old dilution. Disinfectant containers should be washed with boiling water after use to kill any resistant bacteria which may have developed.

Mechanical dishwashers disinfect by virtue of the high rinse temperature achieved. Recommended temperatures should ensure that plates, etc should come out clean and too hot to handle. If dishes are washed in a sink they should be rinsed if possible in another sink but always in clean, fresh, very hot water. The use of cloths, unless disposable, should be discouraged as they are an ideal vehicle for transferring contamination. Clean items should be air-dried away from dirty items. Decks do not normally come into contact with food; therefore they do not need to be disinfected.

Food and equipment must not be exposed to contamination during cleaning operations. For example, utensils are often stored in the bottom shelf of an open unit leaving them exposed to contamination from hose water used to clean the deck area.

Ventilation hoods and grease filters should be cleaned regularly. The inside surfaces of ducting should be cleaned at least once every three months. Only trained personnel, using a safe means of access should remove grease filters for cleaning and clean grease and oil from hoods and ducts. Galley crew should be aware of the potential for serious fires in ventilation ducting.

Cleaning schedules are useful, even in a small vessel, to remind staff when specific equipment or areas such as store-rooms require to be cleaned.

11. Pest Control

Good housekeeping obviously minimises the risk of infestation and it is important to ensure that areas, particularly dry stores and refuse areas, are kept in a clean and tidy condition. Lids should always be kept on waste bins which should be washed after emptying.

Electronic fly-killers are effective so long as they are not positioned over food or equipment or in draughts as dead flies may be blown out of the tray. Catch trays should be emptied frequently.

Cockroach presence in galleys and food stores, although less common nowadays, in some cases still persists, (Figure 4). During the day they hide in dark corners, cracks, pipe ducts etc. Several types of traps or sticky pads are available to monitor the extent of infestation. They should be numbered and checked regularly. The number of cockroaches in each trap should be duly recorded and charted. Regular monitoring should continue well beyond treatment to determine whether any cockroaches have survived. To obtain satisfactory results, any treatment must break the life cycle. This may involve a period of 12 weeks which

Figure 4. Cockroach infestations still persist on some ships

is the time taken for eggs to develop. Insecticides generally fall into two categories, ie quick-acting (with flushing agent) and residual. Ideally a quick-acting type insecticide should be used and followed up with a residual type application which remains active over a prolonged period. Changes in technology have produced a number of more environmentally friendly products. For example, some of the 'one-drop' gel products seem to be very effective and relatively simple to administer. Certain products may, however, be restricted and require specialist personnel to administer treatments.

If infestation continues and is clearly beyond the ship's control, action should be taken to secure the services of a specialist pest control company. The use of rodenticides is also strictly controlled. The presence of rats onboard a ship should be reported to the local port health authority who will provide specialist advice and assistance to deal with the problem, (Figure 5). In extreme circumstances, ships may require to be fumigated and this would be carried out according to the IMO publication *Recommendations on the Safe Use of Pesticides in Ships*.

Ships require to have a current de-ratting certificate or exemption certificate issued in an approved port showing that the ship was inspected by the local authority and found free from rodents.

12. Potable Water

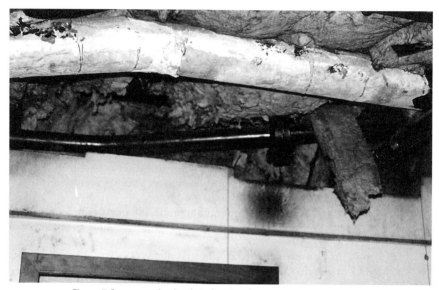

Figure 5. Smear marks clearly indicating the presence of rats onboard

Potable water is, like food, a subject of health concern and of regulation. Specific regulations and actions are required to ensure the quality of potable water and they are dealt with in chapter 5. Potable water should be bright, clear, virtually colourless and aerated, that is it should bubble when shaken. Bright, clear, sparkling water may still however contain harmful levels of bacteria (see chapter 5).

All water used in the galley for food preparation, utensil cleaning and the preparation of detergent and disinfectant solutions should be potable water.

13. Maintaining Standards

Management commitment is essential in order to achieve and maintain good hygiene standards of operation. All food handlers should receive appropriate food hygiene education and training. They should also receive such training as is necessary to ensure they are capable of using equipment, and have an awareness of health and safety aspects within the catering environment. Weekly hygiene inspections on UK-registered ships and other foreign-flag ships of countries ratifying ILO Convention No 68 should be carried out by the master or designated person according to the UK Merchant Shipping

Regulations or other regulations arising from the Convention. Such persons should also receive basic hygiene awareness training to help them understand more than the aspect of general cleanliness. Inspections should follow a logical pattern, from receipt through storage, preparation, service and eventual disposal.

14. Food Poisoning (Ref: The Ship Captain's Medical Guide)

Despite precautions and regulations there may be an outbreak of food poisoning or illness with diarrhoea on board. The following steps should be taken:

- Inform the relevant port health authority of the circumstances according to the International Health Regulations and Public Health (Ships) Regulations 1979.
- Try to identify the food(s) which have caused the outbreak. Make sure that no further consumption of these foods is possible by discarding them. Keep a sample, if near port, so that laboratory identification is possible.
- Inspect all food handlers for general cleanliness and apparent standard of health, including septic skin conditions.
- Check if anyone has suffered from a communicable disease or been in contact with anyone suffering from a communicable disease such as dysentery or typhoid fever.
- Has anyone had an unexplained or unusual illness?
- Exclude from catering facilities anyone who could be considered a potential disease carrier.
- Inspect all food storage and preparation areas, and make sure that good food hygiene practices are being followed.
- If near port, keep samples of diarrhoea or vomit (in sealed containers inside a plastic bag in a cold place away from catering facilities) so that laboratory identification is possible.
- Treat all affected persons and take whatever precautions are necessary to prevent the spread of the disease.

15. Appendix

Maritime and Coastguard Agency responsibilities

The UK Maritime and Coastguard Agency (MCA) responsibilities for food and catering for crews on board ship arise from the 1946 International Labour Organisation Convention No 68. The Convention requires ships to have onboard food and water supplies which, having regard to the size of the crew and the duration and nature of the voyage, are suitable in respect of quantity, nutritive value, quality and variety. It further requires the arrangement and equipment of the catering department in every ship permit the service of proper meals to members of crew.

These responsibilities are implemented for the UK principally through the Merchant Shipping (Provisions and Water) Regulations 1989, the Merchant Shipping (Crew Accommodation) Regulations 1997 and the Merchant Shipping (Crew Accommodation) Fishing Vessel Regulations 1978. It is a requirement of the MS (Provisions and Water) Regulations that masters regularly inspect the accommodation and facilities, including the water supply.

Developments in food technology and emphasis on quality management over the past few years has led to a change from the traditional 'pork and beans' stereotype to a more focused approach to inspections consistent with current trends throughout the catering industry and in accord with The Food Safety Act and the Food Hygiene Regulations, albeit the regulations mainly apply to shore-side premises. The new regulations have had a major impact within the catering industry, focusing on how to identify and control food safety risks at each stage of the process from point of delivery to point of service of prepared food, and on the importance of training food handlers to a level commensurate with their job.

Merchant Shipping Notices M1373 and M1375 have been superseded by Guidelines for Food Hygiene on Merchant Ships and Fishing Vessels (MGN61 1998). The inspection of food handling areas form part of the overall ship inspection and MCA marine surveyors carry out inspections based on fundamental rules of hygiene. The MCA food and hygiene inspector has overall responsibility for food and hygiene matters and visits marine offices throughout the UK to provide specialist advice and support to local marine surveyors during ship inspections.

Port health officers carry out shipboard inspections, issue derat certificates and exercise their statutory public health powers which may overlap MCA's responsibilities. A memorandum of understanding between respective organisations ensures a close liaison is maintained between respective authorities so that where inspections carried out by members of one organisation reveal a situation which will obviously be of concern to the other, appropriate action under their respective powers can be considered. A simple explanation of respective roles may be that port health officers are primarily concerned with the health of the nation, protecting the mainland and preventing disease from entering the country. Maritime and Coastguard Agency surveyors, in the context of this chapter, are concerned with protecting the conditions of the seafarer.

8.   Microbes in Cargoes

D Robbins

Contents

1. Introduction

It is a common experience that produce of plant and animal origin, left in the field, rots and decays through the action of microbes. If conditions are suitable, the same organisms can grow in the materials after harvest, processing or refinement, causing them to be spoiled. These materials comprise a large proportion of the cargoes carried at sea, in addition to much of the packaging for other inert goods. Conditions favourable to the growth of microbes in a cargo or hold are clearly unfavourable to maintaining its quality during carriage and reduce the likelihood of it arriving in a sound state.

As well as vegetable or animal produce, there are other organic and inorganic cargoes, such as petroleum oils and sulphur, which are possibly less widely known to be capable of supporting microbial populations, and significant problems can arise during carriage of such cargo.

Although the carrier is usually entitled to certain immunities, national and international insurance and rules of carriage, such as the Hague-Visby Rules, usually include wording to the effect that the carrier

'...shall properly load, handle, stow, carry, keep, care for and discharge the goods carried'.

Owing to the wide variety of goods carried at sea and the very large number of possible spoilage organisms, it is not feasible to describe every instance of spoilage in a short chapter. This chapter seeks to provide the technical background behind the causes of microbial spoilage in cargoes, its effects and the strategies to assist the carrier minimise damage occurring at sea. It is illustrated with some examples of common cargoes.

2. The Effects Of Microbial Contamination

2.1 Staining and Discoloration

To a buyer, the appearance of commodities is often a consideration as important as function and is sometimes the sole reason for purchase. Although microbial contamination may not impair a product's function, an adverse change in its appearance, or its packaging, could cause it to be rejected. This will be the case with highly processed goods, decorative items and foodstuffs.

Surface discoloration - commonly black, white or green, but also pink and yellow - is the most readily recognisable sign of microbial contamination. This discoloration on solid goods is generally caused by the growth of moulds that produce coloured spores or are themselves coloured; for example, the green spores of *Penicillium* mould species and black or yellow spores of some *Aspergillus* moulds, which appear on vegetable matter, including paper and carton board.

The hyphae (the vegetative growing shoots) of some moulds are coloured, such as those of *Ceratocystis*, a mould that causes blue stain in wood. Crystalline or resinous pigments, secreted by moulds, have a low solubility and, as mould growth is invasive, the staining becomes embedded in the infected material and is virtually permanent. Remedial treatments, such as bleaching, are unlikely to be successful and bleaching chemicals can cause their own problems on sensitive materials.

Lumber is colonised by blue stain (sap stain) mould soon after felling. Whole logs dry relatively slowly, so significant discoloration can occur before sawing, and wood is one of the few raw materials that is treated with a preservative, usually a water-soluble fungicide, prior to storage and transport.

Tree bark is an effective barrier against colonisation and if lumber is carried with the bark in place, care should be taken to minimise damage to it. Although stain mould growth will not significantly affect the strength of the wood, it will almost inevitably lead to a loss of value.

In order to 'digest' the material upon which they are growing, microbes secrete enzymes that attack and decompose it. These enzymes also cause colour changes by attacking dyes, especially natural vegetable dyes, and may result in permanent fading or discoloration, an expensive hazard with fabric and textiles.

The early stages of contamination of cereal grains by spoilage fungi leads to dulling of their normal, bright appearance. Although the grain may still be intact at that stage, the dull appearance can be sufficient to cause a reduction in grade quality and, hence, its value. As spoilage advances, the outer layers become darker, followed by discoloration of the germ and endosperm, which eventually turn dark brown or black. The grain will have become unfit for consumption long before this time and it will probably be displaying other signs of deterioration, such as clumping, heating, visible mould growth, and a strong off-odour.

2.2   Self-Heating

The respiratory processes of microbes produce heat, but if infected materials are left in the open this heat goes unnoticed. However, in a large mass of infected material, such as a hold filled with moist grain, heat may be produced faster than it can dissipate, resulting in a rise in temperature of the material.

Moisture in a bulk cargo of material of biological origin will not be distributed evenly, and microbial growth will be most vigorous in the areas of highest moisture. Consequently, microbial heating in a susceptible cargo is generally uneven, possibly occurring in more than one location.

Temperature rises can be modest, in the order of a few degrees, but can often climb to 50°C to 55°C; in the presence of thermophilic organisms, they may rise to around 70°C. As heat increases the rate of most chemical reactions, these elevated temperatures will accelerate loss of quality above that caused directly by the spoilage microbes themselves. Proteins begin to denature, carbohydrates, especially sugars, start to darken, and the free fatty-acid content of oils and fats may increase, producing an increased risk of rancidity.

Vigorous microbial self-heating in a portion of cargo will tend to be self-limiting, as the viability of the microbes generating the heat will be reduced by a rapid rise in temperature. Once established though, the microbial self-heat-

131

ing process is likely to continue. The moisture level will increase in adjacent areas as microbes produce water during respiration and due to moisture migration (see section 4.2), resulting in proliferation of organisms already present and in the invasion by organisms from the heating material. The zone of microbial heating may thus migrate through contaminated cargo over a period of days.

Microbial self-heating may cause concerns other than a loss of quality to the cargo. The oils in some animal and vegetable cargo (see Table 1) react chemically with oxygen in air and are capable of producing temperatures well above those at which microbes are killed, and which may lead to ignition. Experiments with soya beans have shown that microbial and chemical self-heating are separate processes and that the latter may have a threshold temperature below which it can not proceed (indicated to be above 50°C for soya beans). Microbial heating can therefore raise the temperature of a susceptible cargo above the lower threshold at which chemical self-heating is sustained.

The concentration of oxidisable oil present and the availability of air, limit the chemical self-heating process in susceptible material. The concentration of oil

Substance, material or article	UN No.
CARBON animal or vegetable origin	1361
COPRA	1363
COTTON WASTE, OILY	1364
COTTON, WET	1365
FABRICS, ANIMALS with oil; VEGETABLE with oil; FIBRES, ANIMAL with oil; VEGETABLE with oil	1373
FIBRES ANIMAL burnt, wet or damp; VEGETABLE burnt, wet or damp	1372
FISHMEAL, UNSTABILIZED, FISHSCRAP, UNSTABILIZED	1374
Meal, oily; Oil Cake; SEED CAKE, containing vegetable oil;Seed Expellers, Oily	1386
PAPER, UNSATURATED OIL TREATED incompletely dried	1379
SEED CAKE with not more than 1.5% oil and not more than 11% moisture	2217
WOOL WASTE, WET	1387

Table 1. Produce of Animal or Vegetable Origin listed as materials liable to spontaneous combustion (Class 4.2) listed in the IMO IMDG Code 2001. Seed cake includes meal, cake, flakes, expellers and pellets of: coconut (copra), cottonseed, groundnuts, linseed, maize (hominy chomp), niger seed, palm kernel, rape seed, rice bran, soybean, and sunflower seed.

in solvent-extracted oilseed meal is often too low to support vigorous chemical self-heating, and it is doubtful whether spontaneous ignition can occur, even with advanced microbial heating.

The expellers from mechanical processing, however, contain significantly higher concentrations of oil, and particular attention should be paid to the temperature and moisture content of these products upon loading.

Fishmeal is widely known to be capable of self-heating to ignition, and the process is sufficiently hazardous for the IMO, through the International Maritime Dangerous Goods (IMDG) Code, to require that fishmeal be treated with an anti-oxidant prior to loading in order to minimise self-heating during carriage. Since the respiration of growing microbes will tend to exhaust the anti-oxidant, a risk of hazardous self-heating may arise if the fishmeal becomes moist or wet.

Moist cereal grain is known to self-heat as a result of microbial activity, as well as from respiration of the seed itself, and steady temperatures in the region of 50°C to 55°C may be reached. Since cereal grains contain little fatty material they are not considered to pose a risk of self-ignition.

Microbial heating often leads to the appearance of steam from infected cargo, and it should not be mistaken for smoke from combustion. Clearly, careful consideration should be given to the significance of the appearance of steam, and signs of heating in cargo should not be dismissed merely as a microbial problem of no consequence to safety, particularly in those cargoes that are known to self-heat. Standard cargo references, such as the *IMDG Code*, *Thomas's Stowage* or *Lloyd's Survey Handbook*, should be consulted when loading these materials and the advice followed appropriately. If there is doubt, expert help should be sought.

2.3   Off Specification

Microbial spoilage can give rise to a wide range of problems, and some of the symptoms (and reasons for loss of quality) are highlighted below.

In foodstuffs, microbial spoilage can cause the material to be down-graded or declared unfit, for either human or animal consumption, for a number of reasons. Pathogenic organisms, such as Salmonella bacteria, may be present, or microbes may have produced toxins, such as aflatoxin produced by *Aspergillus flavus*, in animal feeds. Microbial acids and, as mentioned in the preceding section, heating, can result in a reduction in the content of key nutrients, which may be detected in standard quality tests carried out upon arrival. Off-odours and off-tastes (tainting), reduced germination of seeds, caking and reduced flowability are all symptoms of microbial contamination. Frequently, heavy microbial spoilage will be associated with infestation by insects.

The metabolic by-products of microbes growing in water associated with oils can lead to increased miscibility of the oil, causing permanent entrainment of organisms, inorganic particles and water in the oil. These will produce symptoms such as increased turbidity and elevated particle counts, and advanced infections can lead to blockage of filters and gauges. If the infection is long standing, microbial hydrogen sulphide may also be produced, which can have serious health and corrosion effects.

Owing to previous instances of severe microbial contamination, contracts of sale for middle distillate fuels now commonly include specifications for maximum numbers of microbes (see Case History, chapter 2, section 6.1). Although the limits chosen are stringent, they do not necessarily bear a direct correlation to poor performance of the oil; they are achievable in a clean product left to settle. An adverse change in the properties of the oil might be suspected if, after settling, microbial numbers remain above the specified limits.

Other symptoms of microbial contamination include the production of slime or rot on meat and fruits and the production of gas, leading to over-pressurisation and, possibly, to bursting of containers

2.4 Corrosion

In dry cargoes, microbial activity has little effect on the structure of the ship. However, microbes will proliferate in water associated with an oil cargo (product or crude oil), or that draining from solid cargo. The organic acids, and some inorganic acids, produced by microbial films and sludge on the tank top and other surfaces, can contribute to corrosion damage of exposed steelwork.

Depending on the cleanliness of the upstream delivery chain, a parcel of oil could contain a significant amount of water in suspension when loaded, that will settle out in time on the tank top. Any organisms present in the oil will be associated with this water and will thus settle out with it. In this and other stagnant (anaerobic) conditions, significant numbers of sulphide generating bacteria will, almost inevitably appear. As discussed in chapter 1 of this book, these organisms can result in aggressive corrosion, and average corrosion rates of $220mg/dm^2/day$ have been measured in association with the growth of sulphide generating bacteria on mild steel.

Besides producing corrosive metabolites, microbes can contribute to corrosion processes in other ways: by forming differential aeration and concentration cells; by depolarisation of cathodic processes; by disrupting natural protective films and by damaging protective coatings.

If the coating of a tank or hold is maintained and is intact, it is unlikely that there will be any significant corrosion effects from microbial growth, even when sulphide generating bacteria are present. However, the coatings themselves have varying resistance to microbial attack.

3. Conditions Affecting the Growth of Microbes in a Cargo

3.1 An Inoculum of Microbes

In order for microbial spoilage to occur, a material will have to be contaminated with microbes that can grow on or in it so as to cause adverse changes. In most instances the spoilage organisms are those associated with natural decay and, together with their spores, they are found on the produce in the field. They will be present in process plants, especially where water-based processing fluids are used, in storage silos, holding yards or even individual transport vehicles. It is, therefore, unrealistic to expect all but highly processed and specially packaged cargo to be entirely free of spoilage organisms. Should the required conditions for growth subsequently prevail, the spoilage organisms will proliferate.

Cargo debris and loose scale on a ship may act as reservoirs for contaminants, especially if they remain moist. Large numbers of organisms can be present in films and sediments remaining in the bilges, which will contaminate cargo if they overflow. Sound housekeeping practices can reduce the chances of a cargo receiving a large inoculum of spoilage organisms from the ship. Holds should be clean, dry and well-ventilated before receiving cargo and the bilges kept as clean and dry as practical and properly sealed when the cargo dictates. The main hatches and access hatches should be weathertight.

When carrying susceptible liquid cargo, such as mineral and vegetable oils, the loading lines should be thoroughly flushed and allowed to drain, as should any legs of pipe, such as heater bypasses or vent lines that might be used for additional loading capacity. Clean oil loaded into a clean, dry tank, through clean, dry lines, should not deteriorate during a voyage, although this has not prevented suspicion falling on a ship when the outrun of oil has been found to be of poor quality. If full loading and unloading samples are held, it may be possible, by submitting the samples for expert analysis, to determine whether the vessel was likely to have contributed to the problem.

3.2   Nutrients

Virtually every natural, organic material that is not treated with a preservative is capable of supporting the growth of microbes, and the availability of nutrients is not, in practice, a limiting factor in the potential for microbial spoilage.

Some microbes have become so adapted that they can derive all their nutritional and energy requirements from inorganic materials, in darkness. Sulphur and iron pyrites are two inorganic materials that some specialised organisms (species of *Thiobacillus* bacteria), can utilise. Cargoes of iron ore and sulphur usually contain free water, and significant populations of *Thiobacilli* can develop. These organisms are particularly important as they produce and continue to grow in highly acidic conditions, as low as pH2, which will clearly exacerbate corrosion of exposed steel. They have also been reported as present in water associated with cargoes of high sulphur coal. When carrying these cargoes, close attention should be paid to keeping the bilges pumped dry at all times. Holds in which sulphur is carried are lime washed and should be inspected before loading to ensure that the walls are properly coated.

3.3   A Suitable Chemical Environment

Different groups of microbe require different chemical environments. As well as thriving at low pH, some can withstand alkaline (high pH) conditions. Microbes commonly grow in the presence of oxygen (aerobic conditions), but there are many important spoilage organisms, such as the sulphide generating bacteria, that thrive in anaerobic conditions. Some microbes can also withstand high salt concentrations (halophilic organisms), or tolerate and accumulate toxic heavy metals.

The range of chemical environments that microbes can exploit is as broad as the conditions that arise in organic cargoes and, providing adequate moisture is present, it is seldom the case that the chemical environment in a cargo will prevent microbial spoilage.

3.4   A Suitable Temperature Range

Microbes tolerate a wide range of temperatures, from below freezing to the boiling temperatures of hot, sulphur springs. Microbes may be divided into three categories according to the temperature range in which they prefer (- phile) to grow:

Psychrophiles	-3°C to +15°C
Mesophiles	13°C to 45°C
Thermophiles	42°C to 100°C

The optimum temperatures for each group tend toward the middle of the range, and the limits for possible growth in each category overlap and are somewhat arbitrary. The majority of microbes are mesophiles and these are the commonest spoilage organisms, although organisms living towards the extremes of high and low temperatures are also involved in the microbial spoilage of cargoes.

Black Spot mould (*Cladisporium herbarum*) grows on refrigerated meats and has been reported as a contaminant on beef carried to Europe from South America, growing at less than -5°C. Other microbes, such as the mould-like Mucor, which can grow at freezing point, have also been found on chilled meats. Growth tends to occur where there has been surface condensation or temperatures have fluctuated above zero, for example near electric lights. Thus, when loading in humid climates, meat should not be left exposed, and close attention should be paid to carcasses which show evidence of condensation or being incompletely frozen.

Although thermophilic microbes, such as the mould *Aspergillus flavus*, are not generally those that first initiate spoilage, they increase in prominence after other, mesophilic microbes have proliferated and caused the temperature of the cargo to rise above ambient.

Thermophiles can cause the temperature of a cargo, or a heavily contaminated portion of it, to rise above 70°C, resulting in adverse changes in most natural materials or leading to other hazards (see section 2.2).

The most significant effect that temperature has on microbial spoilage of cargoes is its strong effect on humidity.

3.5   Moisture

All living organisms require water to grow, and drying is probably the oldest known method of prolonging storage and ensuring safe transport of goods. Microbial spores are, however, resistant to desiccation and a cargo that was loaded in sound condition can deteriorate if it becomes wet or if it absorbs sufficient moisture from the atmosphere. Controlling the moisture present in a cargo space is therefore important in ensuring that perishable goods arrive at their destination free from microbial spoilage. The relationship between the moisture content of a cargo and the air surrounding it are considered in more detail below.

Organic materials, which includes most cargo and packaging, are to a lesser or greater extent hygroscopic, meaning that they can absorb or desorb atmospheric moisture. If they are drier than the surrounding air they will absorb moisture, and will give up moisture if the air is drier. The transfer of moisture between a hygroscopic cargo and the atmosphere stops when the relative humidities within the material and the surrounding air are equal. At this point the material is said to have reached its equilibrium moisture content with the air.

The moisture content of agricultural cargo is usually expressed as a percentage of its wet weight, though some scientific laboratories may report measurements as a percentage of a cargo's dry weight. To convert one to the other refer to Appendix I. Atmospheric moisture is expressed as percent relative humidity:

$$\%RH = 100.h/h_o$$

Where h = specific humidity (g water/kg dry air), and h_o = specific humidity at saturation

Of the three groups of microbes - bacteria, moulds and yeasts - bacteria have the highest requirement for water, needing virtually saturated conditions, almost 100% relative humidity, to grow. Yeasts also require %RH values close to 100. Moulds, though, can grow at lower moisture concentrations, as low as 65%RH and are consequently the commonest and most prolific spoilage organisms. Each type of mould will have a minimum moisture requirement (see Table 2).

Cargo with a moisture content that has an equilibrium relative humidity of 65%RH or greater is at risk of microbial spoilage although growth rates at

Mould species or group	Starchy grains	Soybeans	Peanuts and copra
Aspergillus restrictus[1]	14.0 - 14.5	12.0 - 12.5	8.5 - 9.0
A. glaucus	14.5 - 15.0	12.5 - 13.0	9.0 - 9.5
A. candidus	15.5 - 16.0	14.5 - 15.0	9.0 - 9.5
A. flavus	17.0 - 18.0	17.0 - 17.5	10.0 - 10.5
Penicillium	16.5 - 20.0	17.0 - 20.0	10.0 - 15.0
Fusarium	22	-	-

[1] A. restrictus grows in the range of approximately 68%RH - 73%RH.

Table 2. Approximate lower limits of %moisture (wet basis) required for growth of some spoilage moulds on cereals.

65%RH are usually low and for significant mould growth in cargo a minimum range of 70 to 75%RH is generally required. No microbes can grow below 60%RH.

It is therefore helpful to be aware of the equilibrium moisture level for a cargo at 65 to 75%RH, as this will give some indication of the vulnerability of the consignment if a change in moisture content is anticipated. Unfortunately, such information is rarely readily available, though the equilibrium moisture curves for a variety of produce is shown in Figure 1. The moisture contents at 65%RH vary considerably over the range of products and it should also be noted that the vertical axis is moisture content on a dry weight basis, at 24°C.

The concentration of water vapour that can be held in air, and thus potentially absorbed by cargo, is strongly influenced by temperature. Warm air holds more

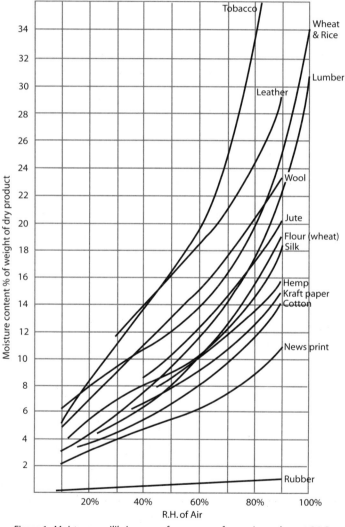

Figure 1. Moisture equilibria curves for a range of organic produce at 24°C.
From Modern Ship Stowage, J. Leeming

water vapour than cool air and a useful approximation is that, in the range of commonly experienced ambient temperature, the concentration of water vapour in saturated air (h_o) doubles for every 10°C rise in temperature (see Table 3).

It can be seen from Table 3 that if saturated air at 20°C is heated by just 5°C, the relative humidity falls to 70%RH (14/20 x 100) even though it still contains

Temperature (°C)	Approximate value of h_o ($g_{water}/kg_{dry\ air}$)
10	7.5
15	11
20	14
25	20
30	28
35	38

Table 3. Specific Humidity h_o at various temperatures.

14g water/kg air. If the same air is cooled to 15°C, the dew-point is passed and approximately 3g of liquid water/kg air will condense.

Changes in relative humidity with temperature can be followed on a psychrometric chart, Figure 2 being an abbreviated version. The point can be found on the chart for air at a known temperature and relative humidity. On cooling, the point moves horizontally to the left and the relative humidity is seen to

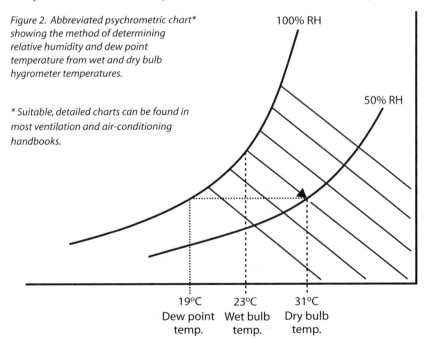

Figure 2. Abbreviated psychrometric chart* showing the method of determining relative humidity and dew point temperature from wet and dry bulb hygrometer temperatures.

* Suitable, detailed charts can be found in most ventilation and air-conditioning handbooks.

100% RH

50% RH

19°C
Dew point temp.

23°C
Wet bulb temp.

31°C
Dry bulb temp.

increase, until it reaches the 100%RH curve. The air has then reached its dew-point temperature, which is read from the horizontal axis directly below. If cooling continues below this temperature water condenses.

The form in which the liquid water appears, depends on the cause of the cooling. If the temperature of the body of air as a whole falls below the dew-point, water appears as a fog or a mist. If the air is cooled because it comes into contact with a cold surface, drops of liquid will condense on that surface.

There are two principal mechanisms by which condensation can occur in a hold. Firstly, surface condensation will form when warm, moist air, thrown off by cargo, comes into contact with parts of the ship's structure that are below the dew-point of the humid air. This is known as ship's sweat because of the appearance of water on the ship's structure and can occur when a vessel has loaded in a warm, tropical port and sails into cooler, temperate climates. As a rule, ship's sweat is prevented by ventilating a hold through the voyage into cooler waters.

The second means by which condensation can occur is when warm, moist air, introduced by ventilation, or produced in heated portions of the cargo, comes into contact with cold cargo. This is known as cargo sweat and occurs when the cargo temperature is below the dew-point of air in the hold. Cargo sweat can occur in a vessel loaded in winter, in a temperate port and which then sails to tropical climes. Cargo sweat is prevented by ensuring that the holds and ventilators stay closed until the cargo has warmed up above the dew-point of the ambient air.

In practice, the large mass of cargo in a closed hold is unlikely to reach ambient temperatures when sailing from a northern port to a tropical destination. When the hatches are opened for discharge, condensation may form on the surface of the cargo. However, once unloaded, the cargo usually soon reaches ambient temperatures and this surface condensation rarely results in significant deterioration. Cargo sweat can be avoided by conditioning the cargo in a heated store prior to loading, or ensuring that the main hatches, access hatches and ventilators remained well sealed during the voyage.

Temperature influences the moisture content of a cargo because of its effect on relative humidity. The majority of ships have no humidity control and the passage of a vessel through various climates on a voyage can therefore result in dramatic changes of humidity in a hold and, as a consequence, the moisture content of the cargo within it. The moisture content of a cargo, the relative humidity of air in the hold and temperature differences provide the basis for determining the likelihood of microbial spoilage occurring during a voyage.

4. Conditions Giving Rise To Spoilage

4.1 The Initial Moisture Content of the Cargo

There is little influence that a carrier can exercise over a bulk cargo, and its temperature and moisture content at the time of loading will largely determine whether the cargo arrives at the discharge port in a sound condition. If a bulk cargo is loaded with a moisture content much in excess of equilibrium with 65 to 75%RH it is possible that some spoilage will occur on longer voyages. Some idea of the likelihood of spoilage can be gained from agricultural safe storage charts (see Table 4).

Product	Moisture content (% wet wt)	Safe storage time (days)	
		60°F	80°F
Soybeans	14	75	25
	18	12	4
Shelled corn	15	259	109
	20	27	10
Wheat	15	Min. 170	30 - 90
	20	10 - 30	<10
Barley	14	455	110
	17	70	14

Table 4. Safe storage periods of grains at 60°F (15.5°C) and 80°F (27°C) for different moisture concentrations. (Derived from References)

Permissible moisture values stated in contracts or bills of lading are usually those that experience has shown are the maximum at which a bulk cargo can normally be carried safely. Suppliers will clearly wish to provide produce close to the maximum permissible water content and it has been known for producers to spray over-dry product with water, or blend it with produce of a higher moisture content. It is most unlikely that mixing will be thorough, so portions of the cargo could be significantly in excess of the certified moisture value. Variance will also occur when a cargo is a composite from a number of farms or processors.

The moisture content shown on a certificate of quality is usually an average of tests carried out on a number of samples, or a single test on a composite of samples. Given that natural products are not completely uniform, the maximum and minimum moisture contents within a well-mixed cargo can vary appreciably from the average certified value, as can be seen from the expected variance in some cargoes of feed ingredients shown in Table 5. Work carried out with corn revealed that a cargo with a reported average moisture content of 14.8% contained individual kernels containing from 12.7% to 17.5% moisture.

Ingredient	Moisture	Fat	Protein
Barley	14.1	8.0	11.2
Corn	9.9	9.7	7.4
Feather Meal	40.2	29.8	7.0
Fishmeal	24.6	24.9	3.4
Meat & Bone Meal	33.3	17.9	5.7
Poultry Meal	33.5	13.5	4.3
Soy Meal (44%)	9.3	62.3	4.0
Soy Meal (48%)	6.8	56.5	2.2
Wheat	8.8	26.9	17.1

Table 5. Expected coefficients of variance for properties of well mixed animal feed ingredients.

Any perishable cargo that has become wet, due to rain or sea water, or that is visibly stained, should be rejected and only loaded when properly dry.

4.2   Moisture Migration

As discussed in the preceding section, moisture is not distributed evenly in a solid organic product. As water vapour will move from an area of high concentration to an area of lower concentration, moisture should eventually become evenly distributed through the cargo. However, at a constant temperature this process is slow and microbes will proliferate in moist material before the relative humidity has fallen to a safe level. Water produced by microbes during respiration can further promote growth.

Vapour pressures are greater at higher temperatures and so the rate of migration of water vapour will be greater when the temperature gradient is steeper. Microbial heating and changes in climate and the sea temperature will produce temperature differences in a hold.

Temperature gradients produced by boiler rooms, chiller compartments, the pipes associated with these spaces and heated fuel tanks, can all cause adverse moisture effects resulting in microbial growth in their vicinity. Measurements taken in bulk corn, carried from North America to the Far East, showed that the cargo on the heated double bottom tanks reached nearly 55°C. As with this example, it is not always feasible to insulate cargo from a heated surface though, wherever possible, chilled or heated surfaces should be insulated and a channel of 15cm formed between them and the cargo to allow for ventilation.

The rate of moisture migration also depends on the composition of the cargo, as well as temperature differences. Migration through whole grains is comparatively slow as the seed coats are intended to minimise transfer of moisture into and out of the seed and the interstitial spaces are small. Materials with large interstitial spaces, such as cottonseed expellers, will allow a greater rate of moisture migration than, say, soya bean meal.

It is not possible to predict whether moisture migration will result in sufficient increases in moisture content in an individual cargo to result in microbial spoilage. However, it is likely to contribute to adverse moisture effects where steep temperature gradients occur and in more hygroscopic cargoes.

4.3   Water Vapour in Air

Owing to the strong influence of temperature on relative humidity, small temperature differences between the cargo and the ventilating air can have a marked affect on the moisture content of exposed cargo. The temperature of cargo and the dew-point of atmospheric air become the main consideration, as deposition of water vapour, in the form of sweat, poses the greatest hazard to cargo.

Warm, dry, fast-moving air provides the best ventilation. A hold should not be ventilated if the cargo temperature is below the dew-point of ambient air. This will prevent moisture in the air condensing on the surface of the cargo. A safety margin of 2°C to 3°C may be considered prudent so that holds can be sealed before the danger point is reached. The temperature of the ship should also be above dew-point of the ventilating air. Ventilating a cargo loaded in a temperate port while sailing to a tropical port will be likely to result in extensive moisture damage.

The dew-point and relative humidity of air, can be determined using a wet and dry bulb hygrometer and a psychrometric chart. The wet bulb temperature lies on the 100%RH curve and the ambient relative humidity is given by the curve that crosses the intersection of the diagonal line from wet bulb temperature and the vertical line from the dry bulb temperature (see Figure 2). The dew-point can then be determined as described in section 3.5. These values

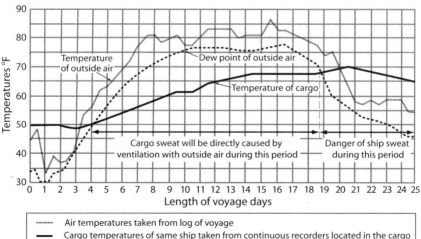

Figure 3. A copy of the record of temperature and dew point of outside air, compared with cargo temperature, for a voyage from Montreal to Vancouver in May 1933. From Modern Ship Stowage, J Leeming

should be tabulated with the cargo temperature, or plotted on a chart, daily, as shown in Figure 3.

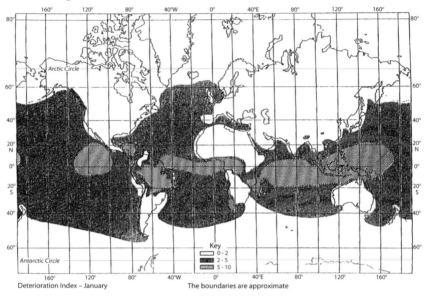

Deterioration Index – January   The boundaries are approximate

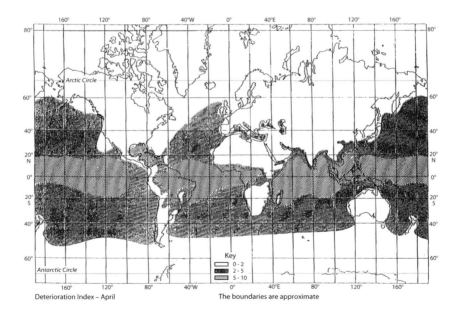

Deterioration Index – April   The boundaries are approximate

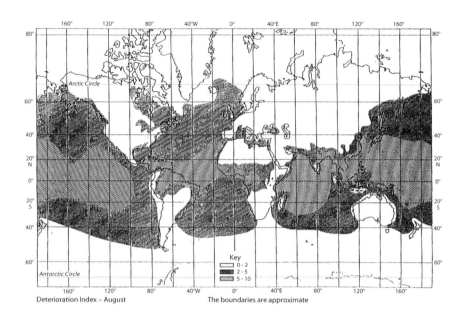

Deterioration Index – August   The boundaries are approximate

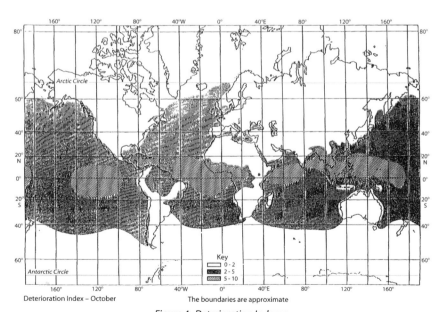

Deterioration Index – October   The boundaries are approximate

Figure 4. Deterioration Indexes

There are situations that can result in sweat forming even when following the ventilation guidelines. A ship carrying bulk cargo, sailing north in winter from the tropics, will ventilate holds with cool air with a lower dew-point than the temperature of the cargo. If ventilation of the holds is vigorous, when the outside air temperature has fallen sharply, exposed cargo will be cooled quickly, possibly to a temperature below the dew-point of warm air rising through the cargo. Thus, although the exposed surface of the cargo may remain visibly dry, moisture will deposit in a layer below it. This may be found as a band of caked, mouldy cargo, probably less than 1m from the surface.

Some idea of the likelihood as to whether atmospheric moisture and temperature changes will cause problems during a voyage can be obtained from considering suitable weather records. A deterioration index (I) was devised by Brooks, as follows:

$$I = \frac{(H - 65) \times V_x}{100}$$

Where H = mean monthly relative humidity and V_x = saturation vapour pressure at the mean monthly temperature.

The deterioration index was calculated using records of mean monthly temperature and relative humidity at sea and the results were plotted on maps. Copies of these maps, from a useful British Standard (*Guide to Hazards in the Transport and Storage of Packages. BS 4672 : Parts 1 & 2 : 1971*) are shown as Figure 4. The index thus provides an indication of the extent to which the prevailing mean relative humidity of ventilating air is likely to be in excess of the minimum at which microbial growth can occur. Values of I = 0 - 2 have a low deterioration index, values of I = 2 - 5 are moderate, and I = 5 - 10 are considered to have a high deterioration index.

Rapid changes in sea temperature also pose a threat of adverse moisture changes in a hold. Charts showing common sea routes along which sea temperatures may change relatively quickly with distance, and common areas for gales and sea fog are also provided in BS 4672 : Part 2 : 1971 (see Figure 5).

4.4   Moisture in Dunnage and Packing

Effective dunnaging of cargo ensures its stability and also reduces the possibility of microbial spoilage by ensuring adequate ventilation. Wood is the most commonly used material for dunnage or for staying cargo and it can contain a significant proportion of water. At high humidities there is a rapid change in moisture content: at 90%RH lumber contains approximately 22% moisture, and over 30% moisture at 100%RH.

As several tons of dunnage might be used in a single hold, a significant amount of free water may be loaded alongside sound cargo. Moreover, a significant proportion of the wood will be present at the sides of a hold where temperatures will be most likely to change. Thus, when receiving fresh dunnage some consideration should be given to its origin. Timber from tropical regions will have

Figure 5. Hazards to goods in transit by Sea

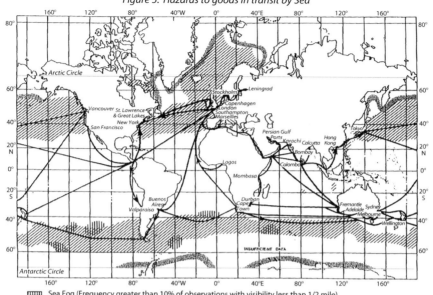

IIIII Sea Fog (Frequency greater than 10% of observations with visibility less than 1/2 mile)

ZZZZ Gales (frequency, of winds Beaufort Force 7 and above, greater than 10% of observations)

Probable limit of Pack Ice – Some principal shipping routes

→ Sections of routes along which sea temperature may fall rapidly in the direction of arrow

Hazards to goods in transit by sea – January

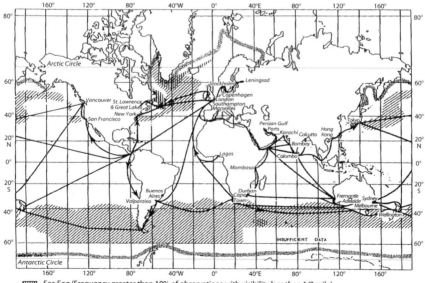

IIIII Sea Fog (Frequency greater than 10% of observations with visibility less than 1/2 mile)

ZZZZ Gales (frequency, of winds Beaufort Force 7 and above, greater than 10% of observations)

Probable limit of Pack Ice – Some principal shipping routes

→ Sections of routes along which sea temperature may fall rapidly in the direction of arrow

Hazards to goods in transit by sea – April

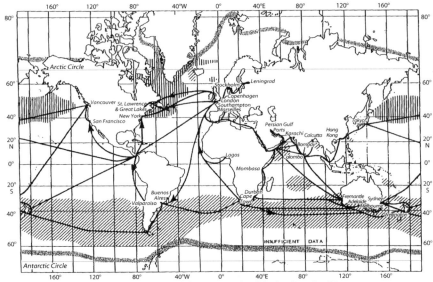

Sea Fog (Frequency greater than 10% of observations with visibility less than 1/2 mile)

Gales (frequency, of winds Beaufort Force 7 and above, greater than 10% of observations)

Probable limit of Pack Ice – Some principal shipping routes

Sections of routes along which sea temperature may fall rapidly in the direction of arrow

Hazards to goods in transit by sea – July

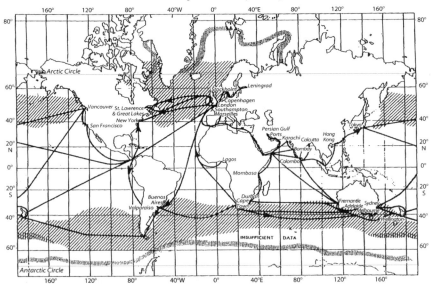

Sea Fog (Frequency greater than 10% of observations with visibility less than 1/2 mile)

Gales (frequency, of winds Beaufort Force 7 and above, greater than 10% of observations)

Probable limit of Pack Ice – Some principal shipping routes

Sections of routes along which sea temperature may fall rapidly in the direction of arrow

Hazards to goods in transit by sea – October

a high moisture content, and freshly-cut wood should ideally be avoided.

Dunnage taken on in temperate ports should also be treated with caution, particularly in winter, as that is the time of year when atmospheric relative humidity is at its highest in these regions. As dunnage is often stored outside, or in unheated stores, it will be at equilibrium with these high humidities and at a comparatively low temperature. If this dunnage is used in a ship that sails into warm water, significant amounts of water may be thrown off as the wood warms up, putting adjacent cargo at risk.

4.5   The Method of Carriage

The significance and nature of microbial spoilage that might occur in a cargo will clearly depend on how moisture changes might occur and this will depend on the form in which the cargo is carried.

4.5.1   Loose bulk cargo.

Temperature and moisture changes in bulk cargo are relatively slow and, as indicated in section 4.1, the majority of the cargo will be unloaded in much the same condition in which it was loaded, providing that the initial moisture content was at a safe level. Moisture changes are mostly confined to the layers at the sides and surface of the cargo, where temperature changes are most likely to occur. If a bulk cargo shows signs of deterioration, the pattern of damage may give an indication of the cause of the spoilage.

Heating and moulding deep within the cargo is a sign that the material was loaded with an initially-high moisture content. Mouldy cargo on the surface, or at the sides of the stow, can indicate that the cargo has become moist due to cargo sweat or moisture migration. Bands of discoloration and moulding over the surface of the stow suggest that ship's sweat has dripped onto the cargo or that hatch covers have leaked.

4.5.2   Bagged cargo

Unlike bulk cargo, with bagged produce there is an opportunity to stow the cargo in such a fashion as to provide adequate ventilation. With some cargo, such as rice, cocoa and coffee, adequate ventilation is essential. Moisture migration can be rapid in these and other bagged cargo, and rapid temperature changes can cause extensive damage if ventilation in the stow is inadequate. Cargoes known to self-heat should also be provided with plentiful ventilation, such as double strip stowage. It is often the case, however, that the method of stowing is a commercial exercise of maximising the quantity of cargo that can be carried in a hold, rather than a practical exercise in optimising ventilation. As such, if spoilage occurs, the pattern of damage may reflect the cause, as with bulk cargo.

4.5.3 Containers

Containers represent small cargo holds and the same principles of moisture migration and moisture changes with temperature apply to containers as they do to cargo holds. The fate of moisture inside a container will depend on the size of the ventilation openings, if there are any at all, and the range of temperatures to which it is exposed.

As containers are relatively small, the whole of the contents will cool down or heat up comparatively quickly, leading to rapid changes in temperature. If there are no ventilation openings, such changes are likely to result in the formation of sweat in the cargo or on the inner surfaces of the container.

Attempts are sometimes made to minimise the effects of moisture changes in unventilated containers by including a desiccant, such as silica gel. In order to be successful, sufficient desiccant must be added to absorb all moisture thrown off, without becoming saturated. If insufficient desiccant is included, it can become saturated and it will then act as a ready source of moisture when the temperature of the container increases.

Containers are often left exposed to the sun's heat in container yards, or on deck, for significant periods of time. When skies are clear, solar radiation at midday can produce surface temperatures the region of 60°C to 70°C on a steel container. Depending on geographical location and season, the temperature can then fall by 50°C or so at night. Such extreme and rapid changes will have dramatic effects on the behaviour of moisture in an uninsulated container, driving it out of the goods during the day and causing it to condense on the container at night. Insulated containers can eliminate the effects of diurnal temperature changes, the internal temperature tending to follow the daily average value.

Large diurnal fluctuations in temperature are most commonly experienced while in port or in container yards. At sea, the speed of the ship, head wind, sea spray and precipitation usually minimise these extreme temperatures, even in tropical and sub-tropical waters. However, maximum temperatures can be approached at slow speeds and with a following wind.

Diurnal weather changes at sea are less dramatic than on land. Relative humidity is comparatively constant and temperature changes are mostly by solar radiation effects on the vessel. The temperature of the vessel remains close to that of the sea.

Moisture is also harmful to other cargoes not subject to microbial spoilage, by causing staining, rusting, caking, germination of seeds/grain (wheat at 25% moisture) and off-specification crushing. Care should therefore be taken if there is mixed cargo in one hold.

5. Safety

Microbial activity in cargo spaces has led to loss of life on numerous occasions due to the production of toxic gases or by depletion of oxygen in the atmosphere.

On the whole, microbes growing in the presence of air consume oxygen and produce CO_2. As virtually all organic cargoes are capable of supporting microbial growth, there is a risk of serious oxygen depletion in an unventilated hold. The normal concentration of oxygen in air is approximately 21% by volume. At concentrations below 17%, human co-ordination becomes poor. Nausea and unconsciousness occur immediately at around 10% and death occurs in about eight minutes. At this point, the victim will be fully aware of the situation but will be unable to cry out or move. Although the hazard of oxygen depletion is well known and well publicised, fatalities still occur when crew enter holds containing natural products, without ventilating beforehand.

In some instances, microbial growth results in the production of toxic gases, most commonly hydrogen sulphide. This gas is produced by bacteria in water containing some organic material under anaerobic conditions. Fatalities from hydrogen sulphide poisoning have occurred when crew have entered apparently gas-free holds to clean them. Thick sludge, comprising rotting cargo, may sometimes remain on a tank top after discharge is complete and hydrogen sulphide gas can be released when it is disturbed.

Rotting cargo can contain pathogenic bacteria, such as *Clostridia*, which are responsible for gas gangrene, tetanus and botulism. Thus, care should be taken to avoid cuts, ingesting the material and forming aerosols when cleaning away the material.

Immediate treatment should be sought for any puncture wounds that are received when working in areas contaminated with rotting or putrefying cargo.

Hydrogen sulphide is more dense than air so it can accumulate in the bottom of tanks and other semi-enclosed low points. The gas has the characteristic odour of bad eggs and can be smelled at concentrations of less than one part per million (1ppm) in air, which gives a good early warning as at concentrations above 300ppm it can cause unconsciousness and death. However, at concentrations of 30ppm to 100ppm, hydrogen sulphide smells sweet, and above this level it anaesthetises the human nose. This is extremely dangerous as it can lead to the belief that the gas has dissipated, when in fact it is present at potentially-lethal concentrations.

Methane gas, which is flammable can also be generated by bacteria and, under some circumstances, could lead to the formation of a flammable atmosphere in an enclosed space.

The correct procedures for entering enclosed spaces should always be followed (see *IMDG Supplement for Bulk Cargo*, Appendix F (IMO) and *Code of Practice for Safe Working of Merchant Seamen* (HSE)), and suitable protective equipment should be issued to crew dealing with significant quantities of rotting cargo.

6.   Appendix I

$$\% \text{ Moisture on dry basis} = 100 \text{ x } \frac{\text{mass product} - \text{mass after drying}}{\text{mass after drying}}$$

$$\% \text{ Moisture on wet basis} = 100 \text{ x } \frac{\text{mass product} - \text{mass after drying}}{\text{mass product}}$$

To convert

$$\% \text{moisture dry basis} = \frac{100 \text{ x } \%\text{moisture wet basis}}{100 - \%\text{moisture wet basis}}$$

and

$$\% \text{moisture wet basis} = \frac{100 \text{ x } \%\text{moisture dry basis}}{100 + \%\text{moisture dry basis}}$$

7.   Acknowledgements

Figures 6 and 9 are reproduced from '*Modern Ship Stowage*', 1968, by Joseph Leeming.

Figure 7 is reproduced from the '*Handbook of Heating, Ventilating and Air Conditioning*' 8th ed., by F. Porges, published by Butterworths, London, by permission of Butterworth Heinemann, Linacre house, Jordan Hill, Oxford, UK.

Figures 11 to 18 are from the '*Guide to Hazards in the Transport and Storage of Packages*'. BS 4672 : Parts 1 and 2 : 1971. Extracts from BS 4672 are reproduced with the permission of BSI under licence number PD\1990 1170. Complete copies of the standard can be obtained by post from BSI Customer Services, 39 Chiswick High Road, London W4 4AL. Readers should note that this standard is obsolescent.

8.   Bibliography

1.   British Standards Institution. *Guide to Hazards in the Transport and Storage of Packages*. BS4672: Parts 1 and 2: 1971.

2.   Christensen, CM (Editor) *Storage of Cereal Grains and their Products*, 3rd Ed. American Association of Cereal Chemists, Inc, St Paul, Minnesota, 1982.

3.   Department of Transport, *Code of Safe Working Practices for Merchant Seamen*, HMSO, London, 1991.

4.   Hill, LD, et al, *Changes in Corn Quality During Export from New Orleans to Japan* Bulletin 788a, University of Illinois, December 1990.

5.   International Maritime Organisation. *International Maritime Dangerous Goods Code, 2001*. IMO, London, October 2000.
6.   Leeming, J. *Modern Ship Stowage* 1968 Printing, with additions, E W Sweetman, New York.
7.   Lower-Hill, BJ, (Editor). *Lloyd's Survey Handbook*, 6th Ed., LLP Ltd, London, 1996.
8.   Milner, M and Geddes, WF. *Grain Storage Studies No. IV*. Cereal Chemistry, September 1946, Vol XXII, No. 5, 449-470.
9.   Milton, RF And Jarrett, K, *The Phenomenon of Moisture Migration in Cargoes of Vegetable Products*, in Carefully To Carry No. 7, The United Kingdom Mutual Steam Ship Assurance Association (Bermuda) Ltd, London, September 1970, 30-47.
10. Rose, AH, (Editor). *Microbial Biodeterioration*. Academic Press, London, 1981.

9. Offshore Oilfield Operations

Dr BN Herbert

Contents

1.   Offshore Oilfield Operations

Both exploration and production phases of an oilfield operation can be adversely affected by unwanted microbial growth. As the presence of free water is a prerequisite for microbial activity, one has only to look to systems where water is, or is likely to be, present to identify those that could be most vulnerable to damaging microbial growth. The range of possibilities is very wide including water-based drilling fluids, completion and hydraulic fluids, hydrotest water, and the crude oil production (both within the reservoir, the wells, the topside production facilities, and pipelines), and also ancillary operations such as cooling systems, power generation, and fuel delivery and storage[1]. Thus some of the microbial problems that are encountered on an offshore oil production facility are covered in other chapters of this book and will not be detailed further. This chapter will focus solely on those systems that pertain directly to oil and gas production. Microbial problems do not seem to feature in reservoirs and facilities that produce gas only (eg southern North Sea) even though water can be present in the produced gas. The use of glycol to dry the gas could result fortuitously in inhibition of microbial growth in the pipelines.

The production of oil and gas from reservoirs located offshore in the central and northern regions of the North Sea started in the 1970s with the discovery of major oilfields such as Forties and Brent. Early platforms were large structures and constructed from steel or concrete. Some of these installations are relatively-simple pumping stations whilst others are complex, with water injection, separation and storage facilities. The need to reduce costs so that oil could be produced economically from smaller reservoirs and those in deeper waters has resulted in the development of new technologies. Thus, satellite wells that are located on the sea bed can be linked to a central fixed production platform or a specially-modified tanker via a series of water injection, production and hydraulic lines that can be several kilometres long. More recent developments have even located separation facilities on the sea bed.

2.   The Microbiology of Oilfields

The reservoir will be anaerobic (oxygen-free). The produced fluids (oil and water) will also be anaerobic but air will be able to enter the production train at various points (eg leaks, separators, sea water ballasting in concrete storage cells) so that the aqueous environment becomes either partially or fully aerobic. The one feature that distinguishes an offshore operation from a land-based one is the presence of sea water. The installations (legs of fixed platforms, sea-bed structures, pipelines) are immersed in aerated sea water. The outer surfaces (both steel and concrete) are subject to the development of a biofilm (see later) of aerobic microbes which is a prerequisite for the subsequent development of macrofouling. This fouling can add substantially to the weight of a structure and could lead to corrosion in the case of steel structures (Figure 1).

*Figure 1. Appearance of corrosion beneath marine fouling on
leg of installation immersed in seawater*

Most of the oilfields in the North Sea require water injection to maintain reservoir energy. The only ready source of sufficient water is the sea surrounding the platform. Aerated sea water is extremely corrosive to the water injection facilities which are constructed mostly from mild steel. Thus oxygen is removed from the sea water to protect the water injectors from aerobic corrosion. This has the additional effect of preventing oxygen entry into the reservoir and hence avoids converting the reservoir environment in the region of the injection wells from anaerobic to aerobic, which could result in an explosion of microbial growth. The biomass produced by aerobic bacteria is usually far greater than that produced by anaerobes. This could be sufficient to cause plugging of the wells, especially if the permeability of the rock is low. However, there are no reported cases of reservoir plugging of North Sea water injection wells - possibly due to rock fractures caused by the injection of cold water into hot reservoirs. It was originally believed that particles found in untreated sea water (eg plankton) could result in reservoir plugging, so it has been common practice to filter the water to a given specification prior to injection. Various filter systems have been used but these can support microbial growth, resulting in plugging of the filters. This has largely been controlled by residual chlorine (where the filter is upstream from the deaerator tower) which is introduced at the sea water intakes to prevent marine fouling. It is now believed by some that fracturing of the injection wells means that unfiltered sea water can be safely injected, so problematical filter systems can be discarded.

2.1   Sulphate Reducing Bacteria (SRB)

The two features that have most influence on the types of microbes that can flourish in offshore oilfield operations are the sulphate in the sea water and the anaerobic conditions. These two provide conditions that support the growth of sulphate reducing bacteria (SRB). It is the generation of H_2S by these obligately anaerobic bacteria (the presence of oxygen prevents their growth) that is perceived to be the main cause of microbially-related problems in oilfields. Reservoir souring (increases in H_2S levels in produced fluids — especially separated gas) and microbially-influenced internal corrosion (MIC) of production facilities have a long history of causing severe losses to the oil industry.

There will be a wide temperature variation throughout an offshore oilfield system. The injected sea water can be initially as low as 5°C, warm up to 15 - 40°C in the treatment facilities, create a temperature gradient in the reservoir (the extent of this from the injection well is controversial) up to the natural reservoir temperature which can be in excess of 100°C, and then progressively cool down again when emerging from the producing wells in the surface separation, storage, and transport facilities down to temperatures of 20°C or below. It was initially thought that SRB could not flourish below 10°C or above 80°C thus restricting activity to topside facilities, the sea water injection well before the temperature rises above 80°C, and the produced oil pipeline prior to the fluids cooling below 10°C. However, there is now evidence of SRB growth at 5°C (long sea water injection lines) and at temperatures in excess of 100°C. The identification of thermophilic SRB in recent years (including the newly identified Archaea) has revolutionised theories on reservoir souring. There are those who now consider that the entire reservoir is at risk from microbial souring but there are still those who believe that souring is mostly restricted to the region of the injection well. Apart from the consideration of possible temperature restriction of SRB activity there is the overlying theory that mixing of sea water with formation water is needed to support substantial SRB activity. The sulphate provided by the sea water needs to be complemented by a source of carbon and nitrogen before substantial SRB activity can occur. Formation waters in most of the North Sea reservoirs contain significant amounts of these nutrients in the form of organic acids (acetate, propionate) and ammonium salts.

One of the most significant developments in recent years has been the realisation that the bulk of microbial activity in the environment occurs on surfaces within deposits that contain biofilm. The biofilm component consists of bacteria and extracellular polysaccharides. This finding has resulted in significant changes in the methods employed for monitoring microbial activity and the assessment of treatment measures (see later).

2.2   Recognition of an SRB Problem

The recognition of either reservoir souring or MIC is, on the face of it, relatively simple. The increase in H_2S levels in produced fluids indicates reservoir souring and the appearance of 'typical' pitting corrosion indicates MIC. However, it is never quite as simple as that. The correct attribution of the observed problem directly to SRB activity can be fraught with difficulty. In the past, the very mention of the presence of SRB would be enough to apply biocide treatment because the fears of the consequences of their activity far outweighed considerations of costs of application, and the emergence of environmental constraints was still in its infancy. Today, the situation is very different. The application of biocides is nowadays, quite correctly, questioned and these are applied only when there is no reasonable doubt that SRB are involved in a problem.

If one considers SRB activity in oilfields as analogous to microbial infection of man, then we rarely apply the rigorous tests that the pathologist employs to diagnose a particular human infection. Additionally, there is usually a long delay (sometimes years) between contamination by SRB and the appearance of the problem which can arise at a location some distance from the site of SRB activity. Perhaps one of the more convincing and pragmatic tests is the application of a biocide resulting in reduced H_2S levels or reduced pitting rates. But this approach can only be truly successful (for diagnosis) if the biocide treatment chosen is effective against the causative SRB and a rapid reduction of suphide levels is observed.

It is now recognised that both souring and MIC are complex phenomena, comprising a range of mechanisms (not all microbially influenced) carried out within consortia of microbes in which SRB may or may not constitute a major part. It is quite understandable that we focus on the SRB to the exclusion of other bacteria because the results of their activity are so visible and they appear (incorrectly) to be easy to detect by the non-specialist. However, because we ignore the other non-SRB members of the microbial community, there are great difficulties in taking representative samples, and the microbiological procedures are not efficient (see later), we obtain only a very partial picture on the microbes present. Even if SRB detection and counting procedures were 100% efficient, it is unlikely that there will be any relationship observed between SRB numbers and H_2S concentrations. It has become, therefore, important that the diagnosis of a microbial problem is derived from a range of data of which the microbiology is only a part.

2.2.1 Reservoir souring:

The diagnosis of SRB involvement can be more clearly established by carrying out a number of tests:
- Confirming the presence of SRB and H_2S.
- Establishing sulphur balances between injected and produced waters. This may be impossible to establish for a number of reasons. Sulphate is difficult to measure, there will be uncertainties of how sea water and formation waters mix in the reservoirs, and the routes of water flow in the reservoir will

probably be unclear (there is rarely any logical pattern to which producing wells and horizons eventually produce H_2S). There can be considerable scavenging of H_2S in the reservoir by Siderite which can result in a complete lack of, or a lengthy delay in, H_2S appearing in the producing wells.

- Demonstrating the consumption of the short-chain organic acids (acetate, propionate etc.) that may be present in the formation water. Their utilisation is evidence of microbial activity.

- Determining the sulphur isotope ratio of the produced sulphide. Sulphur in the environment (eg in the sulphate of sea water) is present as two stable isotopes in a fixed ratio. Sulphide produced by SRB has a different ratio whereas that produced by non-biological processes does not. Thus isotope analysis makes it possible to distinguish between sulphide produced by SRB activity and that produced by a non-microbial mechanism. This procedure is very difficult to carry out and requires expensive equipment and expertise. The results need to be interpreted with care because an unaltered ratio does not necessarily mean that is was a non-microbial process (eg when the sulphate reduction occurs in a sulphate-limiting environment).

2.2.2 Microbially Influenced Corrosion (MIC):

In the case of corrosion, there is no relationship between SRB numbers (whether measured in the water or on surfaces) and corrosion rates. In some instances (eg when a waxy crude oil deposits a hydrophobic layer on the metal surface) there can be large numbers of SRB present but there is no MIC. The tests that are carried out to establish MIC are:

- Confirming the presence of SRB in the water and within deposits.
- Confirming the presence of H_2S in the water is useful but a low or negative result may be misleading if all the H_2S produced has reacted to form iron sulphides in the deposits.
- Establishing the most likely location(s) of the corrosion. This will be obvious if perforation has occurred. In severe cases, penetration of pipeline walls can occur in as little as two months, but several years is more common. MIC can occur on any mild steel surface exposed to water. It follows, therefore, that the lower regions of storage and ballast tanks and crude oil pipelines are at greatest risk because water tends to settle out from crude oil (except when as an untreated emulsion) at flow rates less than 1 ms^{-1}.
- A practical indicator is the detection of iron sulphides in collected deposits by the release of H_2S on acidification. It is this iron sulphide that has the primary role in the aggressive nature of this pitting corrosion.
- Examination of exposed pits is considered of prime importance in diagnosing MIC but needs to be treated with caution. The corrosion is believed to develop as localised pitting that appears typically as a series of concentric rings (Figure 2) that develop under these deposits. In addition, the freshly-exposed metal surface in the bottom of the pit (exposed by removing the

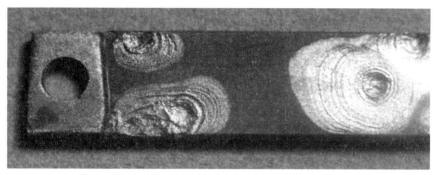

*Figure 2. Corrosion coupons from laboratory rig showing 'typical'
concentric rings associated with MIC caused by SRB*

deposits) appears bright. In practice, however, this is rarely seen and a wide range of pit geometries can develop (Figures 3, 4, 5). There are several recorded instances of long (sometimes several kilometres) grooves in the six o'clock position of cold sea water injection lines that have been attributed to SRB activity. This great variation is probably due to a number of factors including the age of the pitting process at a particular location, the microbial activity, the flow rate, and what other non-microbial corrosion processes are occurring in parallel. The latter is most important to realise because SRB can usually grow in aqueous environments which would be corrosive without microbial activity. For example, these waters may already contain significant levels of CO_2, chlorides, or other corrosive species.

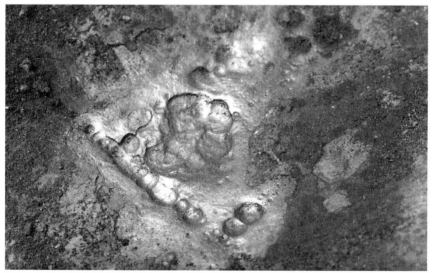

*Figure 3. Corrosion beneath deposits in oil pipeline containing SRB and sulphides. Note shiny
metal surface and line of pitting along weld. Pits have both sharp and smooth edges*

Figure 4. Corrosion beneath deposits in oil pipeline containing SRB and sulphides. Note smooth edges to pits and that metal is not shiny in this case

Figure 5. Corrosion beneath deposits in oil pipeline containing SRB and sulphides. Note sharp edge to large pit and perforation of the metal.

The use of various electrochemical techniques have been examined (linear polarisation, impedance, electrochemical noise, etc) but these have not been proved able to distinguish reliably MIC from other types of corrosion. Usually the recognition of MIC is by default. The corrosion engineer believes that the pitting geometry does not fit their experience for standard corrosion mechanisms, rates of pit development are faster than they would expect from the water chemistry and metal composition, and the presence of SRB has been confirmed by monitoring. It is often on such flimsy evidence that an expensive treatment is initiated.

This increasing uncertainty regarding the specific and quantifiable role of SRB in corrosion had led to a change in its name as views have altered. It was initially called microbial corrosion; this was changed to microbially enhanced corrosion: nowadays it is referred to as microbially influenced corrosion.

3.   Sampling and Evaluation

Procedures for sampling, evaluation, and interpretation of microbial problems in oilfields are given in standard methods published by the American Petroleum Institute[2] and the National Association of Corrosion Engineers[3] and these should be consulted for detailed procedures. Of these, the one published by NACE is the most up-to-date. They include not only the long-established serial dilution method (which depends on the development of black iron sulphide in the culture bottles) for detecting SRB, but also the more recent rapid enzyme methods which depend on the rapid development of a measurable colour (Hydrogenase and APS reductase). The strategies for sampling and testing are invariably devised and implemented by a professionally-competent contractor.

3.1   Reservoir Souring:

Sampling in the reservoir during production is rarely possible. Sometimes, special wells are drilled and/or cores are taken for other reasons which enable in-situ samples to be obtained for analysis. It is from such samples that some workers believe that they have isolated indigenous bacteria from reservoirs (ie not introduced by the drilling or production operations). However, the bulk of the information that we glean on both the chemistry and microbiology of the producing reservoir are from samples obtained from produced fluids and back-flushed injection waters. It is the analysis of these samples that enables us to make some assessment of what is occurring in the reservoir. Lack of detailed knowledge of what is likely to be a very heterogeneous reservoir environment makes such extrapolations of limited value.

3.2   Microbially Influenced Corrosion

The accurate detection of MIC depends on taking samples of deposits at the location where the corrosion is occurring. Most usually, composite deposit samples are obtained from pig scrapings (Figure 6). However, not only do deposits that form in different pipelines vary widely in thickness and composition (having several components including biofilm, scales, sand, corro-

Figure 6. Collection of deposits from a pig that has passed through a pipeline

Figure 7. Example of flush-mounted probe showing arrangement of metal coupons upon which biofilm can develop

sion products and hydrocarbons), there can also be great variations at different locations in a single pipeline. Localised sampling is rarely possible because the sites are inaccessible. In cases of injection or production well tubulars, the samples can only be obtained when the corroded tubulars are recovered (usually after a failure). In the cases of tanks and pipelines, samples can be obtained by the installation, either in the main system or in side-stream devices, of specially-designed, flush-mounted coupons that can double as surfaces on which biofilm can develop, and as corrosion coupons (Figure 7). There are various commercial sources of these. However, these sampling systems can rarely be positioned at the location(s) where the initiation of MIC is thought most likely. In the case of a subsea pipeline, biofilm devices can only be installed at either end, but MIC could be occurring at any point along the line. Even if the device could be installed along the line, the chances of hitting an exact spot where MIC (which is very localised) is going to develop are slim. For these reasons (and also because they need expert handling) such devices have not gained the popularity that was once predicted. Thus, as in the case of tubulars, the best samples are those obtained (after the event) from recovered failed regions of pipelines. It is very important that samples are analysed initially after exposure. Any delay will result in oxidation of corrosion products (Figure 8) and altered microbiology — leading to erroneous conclusions. One should never remove the deposits by grit blasting as this ruins the appearance of the pits - the use of a wire brush usually suffices. Visual examination of exposed pits (a colour photograph is very useful), simple chemistry of collected deposits, and culturing of SRB are all procedures that can be carried out on site.

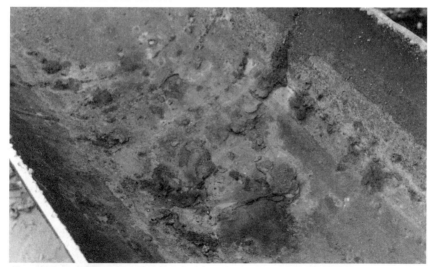

Figure 8. Typical appearance of undisturbed deposits that have formed in an oil-carrying pipeline. On exposure the deposits were black. Fifteen minutes later (time this photograph was taken the sulphides are already oxidising to brown-coloured iron oxides

3.3   Interpretation.

It should be evident from the above that the monitoring data will probably be imprecise — especially the microbiology. In addition, it can be almost impossible to link data from samples to what is occurring at a particular location. Also there is probably (in most cases) no mathematical relationship between SRB numbers and the severity of a problem. The standard methods recognise this and do not give significant levels of SRB. They merely state that the presence of SRB indicates a potential problem. This is not especially helpful because SRB can be isolated from most oilfield operations. The conclusion that a microbial problem is occurring is, therefore, based on consideration of as much additional data as possible, yet, at best will only be qualitative. The most conclusive evidence of MIC is usually obtained at a 'post mortem' of samples obtained from failed metal sections.

4.   Prevention

As water is an unavoidable component of many oilfield operations, the exclusion of this vital vehicle for microbial growth would not seem to be an option in most cases. However, there are many instances where ingress of additional water that has a significant impact on microbial growth could be prevented (eg leaks) by proper construction and maintenance. There are some practices which create almost insoluble problems - eg the use of inaccessible sea water ballasting systems in concrete storage cells. It can be of considerable benefit if a microbiologist could have input at the design stage of a production facility so that the creation of SRB-friendly environments could be avoided where suitable engineering alternatives are possible

Using the monitoring procedures described in standard methods[2, 3] provides a means of being alerted to a problem developing, enabling preventive measures to be applied in good time. Unfortunately, they are rarely applied at the frequency for such an approach to be effective. This is due partly to their unpopularity but also because the results are obtained after long incubation (>ten days). Rapid SRB detection methods (APS reductase, Hydrogenase) have been introduced but have not gained widespread popularity. This is partly due to lack of sensitivity when compared with the traditional cultural techniques.

4.1   Reservoir Souring:

Other possible preventive methods are:
- Replacing sea water for injection by a water source that is free of sulphate. This is normally not possible but it has been suggested that sulphate could be removed from sea water by reverse osmosis. This has not been adopted widely.
- Not discharging separated produced water (rich in organic acids) close to the sea water intakes.

- Not using oxygen scavengers that contain ammonium compounds.
- Using a biocide regime from the very first day of water injection. This is very controversial and is, in effect, not considered usually as an option because of cost and environmental constraints. However, the continuous introduction of chlorine at sea water intakes (to prevent marine fouling) should provide some constraint on subsequent SRB activity upstream of the deaerator towers. Some operators re-inject chlorine after the deaerator towers. The enhancement of chlorine by copper ion generation (pbb concentrations) has been suggested. It has been recommended, by some, that UV sterilisation units be placed within water injection systems to kill bacteria. Although this will reduce (but not completely eliminate) bacterial numbers (already low in clean sea water), it is probable that there will be little benefit because the large volumes of water injected will still introduce significant numbers of SRB.
- New technology is being developed that proposes introducing beneficial microbes into the reservoir that outcompete the SRB or encouraging indigenous competitors. This can be facilitated by changing the nutrient status by injecting chemical packages. Should such an approach prove effective, then parallel biocide use may not be possible.

4.2   Microbially Influenced Corrosion

Other preventative measures are:
- Use of corrosion resistant materials. This requires greater initial investment costs but may prove to be a cheaper option in the long term.
- Use of corrosion inhibitors. The use of some of these products can suffer the same constraints as biocides. The operator should be aware that some corrosion inhibitors (eg amines) can also be effective biocides.

5.   Remedial Action

Remedial action can take a number of forms and is applied after the problem has shown to be in effect.

5.1   Reservoir Souring:

A number of options are possible:
- Shutting in production wells or horizons that yield high levels of H_2S, thereby reducing the overall concentration in produced gas to below the permitted level.
- Include biocides in the injection water. This is the most widely used procedure and for greater details than given here, see Reference 1. Biocides are usually applied intermittently (slug treatment; eg once per week for a fixed period). A range of formulations are provided by a number of suppliers but in reality only a few biocidal molecules are used commonly. There is a range

of formulations based on glutaraldehyde, polymeric biguanide, isothia-zolones and a phosphonium salt — molecules that are claimed to be most effective in controlling SRB activity in reservoirs. Many North Sea operators have moved to a single approved supplier of oilfield chemicals which can reduce the choice of biocides available to them. Most biocide treatments are selected using laboratory tests against planktonic SRB using cultures isolated from the system needing treatment[2, 3]. The results from such tests tend to suggest the application of unrealistically low levels of biocide which prove to be unsuccessful. In more recent years, biofilm tests have been developed[4] which give more realistic data but have not been adopted widely as yet - mostly because they are more complicated. Whatever biocide is chosen, it is important to ensure that it is not more costly than other options, it is compatible with other production chemicals being used, it has suitable partition characteristics in oil/water systems, it has appropriate stability, and any residual biocide can be discharged into the environment within environmental constraints. In many cases biocide application proves to be an unrealistic option.

As for prevention of souring, the application of beneficial organisms and altering the nutrient balance may be an option for treatment.

Probably the most widely used treatment in the North Sea is not to try to stop SRB activity (because this is considered to be an unrealistic option, for the reasons described above) but rather to remove the sulphide from the produced fluids using chemical scavengers. This is usually applied only to the separated gas phase where the limits of H_2S in export gas is far lower than that which would be regarded to be corrosive to the pipelines.

5.2 Microbially Influenced Corrosion

Any treatment that reduces the number of SRB and sulphides in the injection and produced fluids will be of benefit in helping to reduce, or stop, an existing corrosion process. However, when MIC is confirmed, the problem may be too advanced for such a treatment to be effective. Replacement of perforated tubulars, pipelines and other equipment is not uncommon. Such replacement is usually followed by the use of preventive procedures as described above. In some instances both corrosion inhibitor and biocide are used. However, there may be little logic to this rather 'overzealous' approach because an effective corrosion inhibitor should suffice. Biofilms containing corrosion products are almost impossible to treat effectively because biocides are unable to penetrate them. It is usually recommended that any biocide (and for that matter corrosion inhibitor) treatment be complemented with a regular pigging programme to ensure that deposits do not build up. It may be that the selection of a more corrosion-resistant metal will prove to be a sensible option.

6.   Regulatory

For E&P chemicals used on the UK continental shelf, the discharge of chemicals offshore is covered by the Offshore Chemicals Notification Scheme. The scheme is currently voluntary but, at the time of writing, is expected to become statutory in the near future. A testing procedure is based on The Harmonised Chemical Offshore Notification Scheme which was developed by the Oslo & Paris Commission (OSPAR). In the UK, the scheme is administered by the Centre for Environment, Fisheries & Agricultural Science (CEFAS - previously MAFF) on behalf of the Department of Trade & Industry

The maximum amount that is allowed to be discharged without notification (more may be discharged with Ministry permission) in any year is derived from the results of toxicity tests (giving LC50 data - ie concentration that kills 50% of a population) on a range of marine life. The group to which the chemical is assigned is based initially on these toxicity tests but then this may be amended depending on how biodegradable and bioavailable the product is. The relationship between group and notification trigger level from individual installations is shown in Table 1.

Group	Maximum annual discharge without notification (tonnes)
A	40
B	70
C	150
D	375
E	1000

Table 1. Prior notification cumulative tonnage triggers for production chemicals — offshore UK.

It is likely that most biocides will fall into Group A or B but there will be a few exceptions to this. The biocide supplier must register the product so that a group category is available to the purchaser. It should be noted that biocides are considered as part of the total cumulative discharge from an installation. Thus if the trigger level is already reached by other production chemicals, then there is no possibility of using a biocide without prior notification to and agreement from the Ministry.

Information on the regulatory bodies and mechanisms in other European countries can be obtained from the secretary of OSPAR.

7.   Risk Assessment

In recent years there have been attempts to develop mathematical models to predict reservoir souring. Any risk assessment based on incomplete and imprecise data is, at best, difficult. These involve microbiology, water chemistry, hydrology, geochemistry and reservoir engineering. They are helpful in assessing the likelihood of souring at the design concept of a production facility so that suitable materials are used and equipment (eg sulphide scavenging) are provided for. However, these models are unlikely to be refined enough to predict which wells will produce H_2S, the concentration, and when.

In the case of MIC, rudimentary expert systems have been created. These have been kept in-house and have not been published in the open literature. Such systems can only give a likelihood factor of MIC occurring. Nonetheless, they are useful in making informed decisions on whether to use corrosion-resistant materials or install sweetening facilities on the platform. It is much cheaper to install the latter when the platform is constructed than as a retrofit. They are also useful in providing some measure upon which a preventative biocide or corrosion inhibitor treatment could be decided upon.

8.   Safety

A review of safety aspects of offshore microbiology has been published and should be consulted for further information[5]. In general, the risks of infection are very low. It is the microbial products, especially H_2S, which presents the greatest risk to operators. It may be that this safety issue might be of greater significance than souring of products and MIC on particular platforms.

9.   References

1   Herbert, BN, *Biocides in oilfield operations,* Handbook of biocide and preservative use, Ed. HW Rossmoore, 1995, 185-206, Blackie London.
2   American Petroleum Institute, *Recommended practice for biological analysis of subsurface waters,* API RP38, 4th Edn 1990.
3   National Association of Corrosion Engineers, *Field monitoring of bacterial growth in oilfield systems,* NACE standard TMO194-94, Houston USA, 1994.
4   Whitham, TS and Gilbert, PD, *Evaluation of a model biofilm for the ranking of biocide performance,* Journal of Applied Bacteriology, 1993, 75, 529-535.
5   Herbert, BN and Sanders, P, *Survey of microbiological hazards and risks associated with offshore oilfield operations,* Institute of Corrosion technical document, November 1992.